TREASURE UNDER THE TUNDRA

Canada's Arctic Diamonds

L.D. CROSS

VICTORIA · VANCOUVER · CALGARY

Heritage House Publishing Company Ltd.
www.heritagehouse.ca

Library and Archives Canada Cataloguing in Publication
Cross, L. D. (L. Dyan), 1949–
Treasure under the Tundra: Canada's Arctic diamonds / L.D. Cross.

(Amazing stories)
Includes bibliographical references and index.
Issued also in electronic format.
ISBN 978-1-926936-08-6

1. Diamond deposits—Canada, Northern. 2. Diamond mines and mining—Canada, Northern. 3. Mining claims—Canada, Northern. 4. Fipke, C. E. (Charles Edgar). 5. Blusson, Stu. 6. Geologists— Canada. I. Title. II. Series: Amazing stories (Surrey, B.C.)

QE393.C76 2011   553.8'209719   C2011-900353-8

Series editor: Lesley Reynolds.
Proofreader: Liesbeth Leatherbarrow.
Cover design: Chyla Cardinal. Interior design: Frances Hunter.
Cover photo: Mark Evans of HooRoo Graphics/iStockphoto.

The interior of this book was printed on 100% post-consumer recycled paper, processed chlorine free and printed with vegetable-based inks.

Heritage House acknowledges the financial support for its publishing program from the Government of Canada through the Canada Book Fund (CBF), Canada Council for the Arts and the province of British Columbia through the British Columbia Arts Council and the Book Publishing Tax Credit.

  Canadian Patrimoine
Heritage canadien

 Canada Council   Conseil des Arts
for the Arts   du Canada

 BRITISH COLUMBIA
ARTS COUNCIL

14 13 12 11   1 2 3 4 5
Printed in Canada

Contents

PROLOGUE 5

CHAPTER 1 A Diamond is Forever 7

CHAPTER 2 Captain Chaos 14

CHAPTER 3 Dr. Stu 22

CHAPTER 4 Out in the Jungle 29

CHAPTER 5 The Superior Clue 34

CHAPTER 6 Bingo! 51

CHAPTER 7 The Ultimate Find 61

CHAPTER 8 Diamond Stampede 75

CHAPTER 9 New Explorations 88

CHAPTER 10 Tale of Ekati 96

EPILOGUE 104

CANADIAN DIAMOND DISCOVERY TIMELINE 110

GLOSSARY 114

BIBLIOGRAPHY 123

INDEX 124

ACKNOWLEDGEMENTS 126

Prologue

THE THREE MEN STOOD ON *the lakeshore. The water was still blue, but the rust and brown tundra foliage and the skiff of snow on the ground told them it would soon freeze over for the winter. They strode to the south shore and dug up rocks and gravel. Nothing. They quickly moved to the north shore. More digging and rock hammering. Still nothing. Then they moved toward the esker, a huge embankment of sand and gravel snaking across to the horizon. As they passed outcrops of bedrock, one of the men pointed out striations and glacial erratics that indicated ice had passed over this area tens of thousands of years ago. The ground was covered with mud boils heaved out of the ground by frost. Some were over a metre in diameter: big, round pustules with heads full of indicator minerals— just the kind of minerals that indicate the presence of diamonds.*

They went down on their hands and knees in the sand and gravel and filled their pockets with pea-sized green crystals called chrome diopsides. They grabbed smaller stones of red pyrope, a member of the garnet family. The minerals were dispersed all along the width of the shore and up to the esker. They hammered rocks to see what else would be revealed. As geologists, they knew these minerals, when found in glacial sediments, indicate the presence of kimberlite, the kind of rock that could contain diamonds. These minerals had survived glacial transport and were both visually and chemically distinct. So how could they have gone undiscovered for so long?

The men climbed up onto the esker and excitedly looked out over the landscape at the amazing geological history laid out before them. They needed to fill sample bags and test the samples in the lab. They needed to stake their claim to the potential bonanza that lay under the tundra. They needed to prove that this location could churn out millions of dollars of gem- and industrial-grade diamonds like the world had never seen before. Nobody had even believed there could be such treasure under the tundra, and many still remained skeptical. In the cloak-and-dagger world of diamond discovery, they would have to stake out their claim with extreme secrecy. If word got out, their find would be snatched away by every greedy Johnny-come-lately with a plane and a map.

1

A Diamond
is Forever

FIRST CAME JACQUES CARTIER, who set sail from St. Malo, France, to find a passage around or through the New World to Cathay (China and India). If that was not possible, then he was to search for fabulous riches, like the gold the Spanish had found in South America. Cartier believed he had found diamonds and gold at the mouth of Rivière du Cap-Rouge during his third voyage to North America in 1541–42. On his return to France, the samples were discovered to be iron pyrites (fool's gold) and worthless quartz crystals. This episode gave Quebec's Cap Diamant its name and was the basis for the saying "faux comme des diamants du Canada" (as false as diamonds from Canada). Then there was the fool's gold brought back to England by Martin Frobisher in 1576 and his futile attempt to establish a mine for the Cathay Company on Baffin Island in 1578. Next came Samuel Hearne, who made three trips into the "barren lands" west of Hudson Bay tracking down rumours of vast mineral deposits. In 1771, he reached the Coppermine River, north of Great Slave Lake, and the shores of the Arctic Ocean but found only one piece of copper of any value.

The 1896 accidental discovery of placer gold on Rabbit (later Bonanza) Creek, a tributary of the Klondike River, by George Washington Carmack and his Native brothers-in-law Skookum Jim and Tagish Charlie, set off a gold and claim-staking rush in the Yukon. This find was the result of a tip by prospector Robert Henderson, who is now credited as co-discoverer. Today, the treasure is northern ice—not the frozen but the carbon kind.

Diamonds are carbon that has been transformed under extreme heat and pressure into the hardest, clearest mineral on earth. The precise diamond-creating combination of pressure and temperature (44–50 kilobars of pressure at a temperature of 1,000°C) existed a few billion years ago, at depths of 125–200 kilometres (80–125 miles) under stable land. Some volcanoes that originate very deep within the earth force hot magma up to the surface through tubular pipes. This magma carried diamonds up close to the earth's surface before solidifying into rock called kimberlite, a reference to the town of Kimberley in South Africa, where the discovery of an 83.5-carat (16.7-g) diamond in 1871 started a

Kimberlite of Cretaceous age found at Lac de Gras, Northwest Territories. RICHARD HERD. NATURAL RESOURCES CANADA, EARTH SCIENCES SECTOR 2005-128

diamond rush. Only about 30 to 50 of the approximately 6,000 known kimberlite pipes in the world have ever become commercially viable mines. On the tundra, glacial ice covered the volcanic craters, but the movement of these glaciers eroded the kimberlite and carried the debris over the land, giving small clues to the treasure below. Lakes formed in the central depressions of the carrot-shaped kimberlite pipes, making geological exploration and diamond retrieval difficult for all but the most determined searchers.

The word "diamond" comes from the Greek word *adamas*, meaning "indestructible or unconquerable." Diamonds have long been sought after, fought over, prayed to and considered magical, powerful symbols. Their internal fire symbolizes the constant flame of love, and it was said that Cupid's arrows were tipped with diamonds. They have been called shards of stars and tears of the gods. Diamonds can be mined in small-scale hand operations, by tunnelling through rock, placer-sifting through surface gravel, open-pit mining or mining the seabed. Diamonds were known before 3,000 BC, and until their discovery in Brazil in 1725, India was the main supplier.

The De Beers Group cartel (some call it a monopoly) mines 75 percent of the world's diamonds. De Beers is based in Johannesburg, South Africa, and London, England, and has controlled most global diamond production and distribution since its founding in 1888 by Cecil Rhodes. In 1948, De Beers coined the advertising slogan "a diamond is forever," but the catchy phrase was hardly new. The 1925 novel *Gentlemen Prefer Blondes*, by Anita Loos, contained the line: "So I really think that American gentlemen are the best after all, because kissing your hand may make you feel very very good but a diamond and safire [*sic*] bracelet lasts forever."

Many stones obtained by anarchist groups and sold to finance wars, murders and terrorism have been contaminated by the label "blood diamonds" or "conflict diamonds." The Kimberley Process Certification Scheme (KPCS) was established in 2003. It is an international agreement that imposes requirements to certify rough diamonds as conflict-free. Its aim is to prevent the trade in conflict diamonds while helping to protect the legitimate trade in rough diamonds. As of December 2009, the KPCS had 49 members in 75 countries (the European Union counted as one country). The KPCS is a joint government, industry and civil-society initiative. It is not an independent international body but involves a coalition of countries. Because many firms will not buy conflict diamonds directly from their country of origin, the diamonds are often smuggled into and sold through other countries. The KPCS has few tools to deal with corruption among diamond importers beyond asking governments to implement stricter industry regulations. The process relies on self-regulation by the member nations' governments, making it difficult for the KPCS to enforce its own policies. A stronger, independent body composed of both exporters and importers is needed to regulate both sides of the diamond trade.

Loose diamonds can be purchased in their original rough state, but while finding such stones is a big thrill to diamond prospectors, consumers prefer cut and polished diamonds. It is the skill of the miner that brings the stones to the surface, but it is the craftsmanship of the cutter that brings light and life into the gems. Because of their rigid tetrahedral molecular structure (a triangular pyramid with four faces), a diamond can only be cut by another diamond or a laser. Cut diamonds are available in 11 shapes, but the most popular are round (brilliant), cushion, emerald and oval forms. The pear (teardrop) and marquise cuts are often seen in women's jewellery, like necklaces, as are princess and heart-shaped diamonds. The radiant, Asscher (stepped-square) and triangle (trilliant) cuts most often appear in men's jewellery, like cufflinks. Although clear diamonds are traditionally the most highly valued by consumers, diamonds come in different colours, and the more unusual colours can increase market value. Pink and blue diamonds are usually pale, but deep red, green or blue diamonds are prized and fetch high prices. Diamonds are now mined in about 27 countries and on every continent except Europe and Antarctica.

In 1980, there was no such thing as a "Canadian diamond"—everybody knew diamonds only came from Angola, Australia, Botswana, Congo (formerly Zaire), Namibia, Russia and, most important, South Africa. Today, Canada is the world's third-largest producer (by value) of rough diamonds. Canada's reputation as an ethical producer of conflict-free diamonds has made its products economically attractive world wide, but this would never have happened had it not been for the determination and insight of two talented geologists. Prospectors Chuck Fipke and Stu Blusson believed that there were diamonds under the tundra and that they could find them. From 1980 to 1990, they scoured the mountains and the Barrens for samples that would show the indicator minerals often found in association with diamonds. They were adamant that they would find evidence of diamond-bearing kimberlites, those ancient "pipes" of volcanic rock that contain the prized crystals we call diamonds.

2

Captain Chaos

MANY PEOPLE WERE CONVINCED HE was crazy. But Charles Edgar "Chuck" Fipke worked 18 hours a day, 7 days a week in the dark and in the sub-zero cold, and in the end he proved them all wrong.

Chuck was born in Edmonton, Alberta, in 1946 and grew up outside Kelowna, British Columbia, on a rundown 120-hectare (297-acre) backcountry homestead with two younger brothers and a sister. His father, Ed, earned a living taking photographs of people's property. It wasn't a great income, but there was lots of fresh air and freedom. There was also a still, and Ed drank heavily. Chuck wasn't interested in school—he failed English—but he was interested in fighting with his brother Wayne and punched him out at every opportunity. But it was Wayne and his other siblings who would later work with Chuck and invest, often under coercion, in his exploration company. In class, Chuck's quick mind often got ahead of his tongue, making him skip words, stammer and appear dumb. Formal learning was a challenge.

In 1963, when Chuck was 16 years old, he wanted a car to drive into town but needed money to buy one. His father had the solution: Chuck would join the army and learn some discipline too. So that summer he joined the British Columbia Dragoons militia at Camp Vernon. Chuck loved the cannons, guns and the speed and power of the tanks. One afternoon, he and two other new recruits took the tanks out for an impromptu race. They tore out of the camp and roared across the fields, dragging barbed wire behind them. Chuck was in the lead and going down a slope when they came upon a flock of sheep. It was difficult to get a wide field of view from inside the tanks, but the terrified animals were white blurs as they flew out of the way of the teenage tank corps. Inevitably there were casualties. It was difficult to determine who was angrier, the rancher or the camp commander. As punishment, the three offenders were ordered to stand at attention outside all night, ostensibly spotting for forest fires. Since it got a bit chilly at night, Chuck thoughtfully filled their canteens with whisky, just to help keep them warm. The next morning, they were so drunk they couldn't have focused on any fire, near or far. In fact, they could barely stand up. For that transgression, Chuck was assigned the job of radio operator, a boring, indoor task.

By the end of the summer Chuck had earned $125, which was sufficient to buy a 1949 Nash Rambler. His first car was not a hot, classy number, but then he didn't have it for very long. Driving home one day with a buddy, Chuck was challenged to a road race by another driver. Never one to refuse such a proposition, he gunned it, the car skidded sideways, jumped the embankment and rolled several times before coming to rest upright on its worn tires. Both boys were disoriented and had trouble seeing out of the car because the roof had been crushed down near the dash and the seatbacks broken, but the motor was still running. Scrunching down so he could peer out below the new roofline, Chuck managed to drive the car home and parked it beside the orchard near his parents' house. He went out to look for spare parts. Eventually it was joined by other wrecks as Chuck kept working odd jobs and looking for a "real" car.

In Chuck's second try at graduating from grade 13, he had the highest marks at the school in math and science, but English was still a problem. His only reprieve was riding his horse, Cindy, with his shepherd dog, Trixie, at his side. He loved horses, rugby and science. But he was 19, and all of his classmates had moved on. He crammed furiously and just managed to pass English. He hoped that maybe, with his other marks, he could get into university; however, there was another problem, and her name was Marlene Pyett.

Chuck was short and muscular, and his outgoing personality made girls pay attention, but he was most attracted to the dark-haired girl who had stuck with him when his cars had mechanical problems and the bailiff seized Cindy. Marlene had advanced him money to buy another car out of her savings from her job at the bank. But then she had become pregnant, and in the 1960s there were only two choices: adoption or marriage. Her parents paid for wedding invitations, the wedding dress, the church ceremony and the reception. The only thing missing was the groom. Chuck had given it a lot of thought and knew that marriage was an important life event. He made the decision that he would not be married this way. He got into his car and went for a drive.

Not certain that he would be accepted at university and disliking more years of hard study and little money, Chuck was ambivalent about applying to university. A friend of his parents had a seldom-used apartment in Vancouver and a bright yellow Plymouth Fury sport coupe. If Chuck got into university, he could use both. All he had to do was get a student loan and study. Amazingly, he was still friends with Marlene, and with her full support he applied to the University of British Columbia (UBC) and was accepted into general science. He loved birds and considered studying ornithology, but the courses stressed memorizing multitudes of names rather than conducting hands-on research. In spite of their differences, Chuck asked his dad what he thought. Ed suggested geophysics, remembering how that science had been used to find oil under the prairie of Leduc, Alberta, and made the people applying the technology very rich. So, moving from birds to rocks, Chuck enrolled in geology.

In August 1966, Marlene gave birth to a baby boy they called Mark. Mother and baby stayed in Kelowna, and Chuck went to Vancouver, zipping back and forth to campus in the yellow car. At that time, life offered few options for unwed mothers, and Marlene's parents began making arrangements for Mark's adoption. Chuck was against this, but he had no control over his son unless he married Marlene. Just after Christmas 1966, Chuck and Marlene wed in a private ceremony. It was Chuck's choice. He got to keep his son and would have Marlene's unwavering support close by.

Chuck still had not warmed to geology, but in the spring he became intrigued by paleontology and the origins of life on the earth. He was even more enthused when mining companies arrived on campus to hire summer students to work in the mountains near Smithers, an 1,100-kilometre (715-mile) drive north of Vancouver. Maybe these studies would pay off after all. The money and food were good, and they took helicopters when they needed to cut lines or collect samples. At the end of the summer, Marlene and Mark joined Chuck back at UBC in leaky married-student housing. Chuck's marks were better than during his first year, even though he forgot about a major exam because he was absorbed in research in the library. To work off any stress, he followed a strenuous physical regime at the university centre.

The next summer, Chuck left on another job. This time, his travels took him to the Ross River base camp, 200 kilometres (125 miles) from Whitehorse, Yukon, looking for lead and zinc deposits. He loved exploring in the field, and the more remote the better. He worked at Fortin Lake, then in the Selwyn Range of the Mackenzie Mountains, paired up with an old prospector-trapper. Their only contact with the outside world occurred when a helicopter landed to drop off supplies and pick up their samples once every two weeks. By the end of that summer, Chuck had been captured by the North. He wanted to quit school, move up there with Marlene and the baby and live by trapping. Again, it was an older man who gave him valuable advice: "Finish school, work a year and then if you still want to, come back," he said. Chuck went back to UBC.

Chuck got another good job in the summer of 1969, this time out of Norman Wells. The work combined learning, travelling and exploring the Mackenzie River, the lakes that fed it—Athabasca, Great Bear and Great Slave—and the caribou herds and waterfowl that summered on its delta at the Arctic Ocean. The Geological Survey of Canada (GSC) was mapping in a search for hydrocarbon deposits in Arctic basins, and Chuck would work with a team of scientists. It was heady stuff. Maybe he would become an academic. This time he was paired up with a paleontologist looking for trilobites, ancient marine arthropods that could indicate decayed organic matter in ancient seabeds, like the one under the prairie at Leduc, Alberta, where oil had been found.

Chuck and the paleontologist lived in a tent on a mountainside. Chuck mapped the layers of rock, while the trilobite expert recorded the type and frequency of prehistoric marine life. Work done, they broke camp and waited for the helicopter to pick them up. No helicopter arrived, but a storm did. They huddled inside the quickly repitched tent. Then it snowed. Their radio had a limited range, so the next day they trudged through the snow to the mountaintop for better reception. They called out on the radio all day but received no response and had to spend another night in the tent. On the second day, they assessed their meagre supplies and their location at almost 3,000 metres (9,845 feet) on the east slope of the Mackenzie Mountains. It was 150 kilometres (93 miles) to Norman Wells; there was no way it was walkable in the snow. But it was now day three, and the camp had to know they were missing. Day four was bright and clear, but they were still stranded with no rescue craft, a low radio battery, little food and waning strength. Day five came and went. Day six was much better. A helicopter arrived to pick them up, and they were finally off the mountain.

In the spring of 1970, Chuck Fipke graduated from UBC with a bachelor of science in geology (honours). His interest was porphyry copper deposits. He had a multitude of job offers but took none of them. He was accepted into the doctoral program at the University of California at Berkeley and decided he would become a research scientist. He asked a university professor, a former executive with Kennecott Copper Corporation, about the latest technologies to determine ore grades, but he was told that was proprietary information. If Chuck wanted to find out about it, he would have to work for the company. Kennecott Copper did its own cutting-edge research in the field and earned vast profits from its discoveries. Chuck was hooked. Kennecott was in; Berkley was out. Kennecott hired Chuck to find gold and copper in the jungles of New Guinea, and Chuck, Marlene and Mark flew to Australia.

3

Dr. Stu

STEWART LYNN "STU" BLUSSON WAS born in 1939 and raised in the logging town of Powell River, British Columbia, with his older brother, Ross. Their father, a mechanic, drank a lot, and the family had little money. Stu's mother, Edith, bought only what could be paid for in cash.

Stu and Ross often camped out in the bush for days, but they did not shoot anything. Stu hated guns and brutality, so his weapon of choice was a slingshot, which he learned to wield with deadly accuracy. One day he met a hiker returning to town with a pack full of green rocks that contained copper ore. This chance encounter piqued his interest in geology, a career that seemed to offer both money and outdoor adventure. Stu completed an undergraduate degree with top marks at UBC in 1960 and received a doctorate in geology from the University of California at Berkeley in 1964.

At the age of 24, Stu became a permanent GSC employee and was dispatched the old-fashioned way on horseback for a month or more alone in the magnificent mountain wilderness. In the field, he scrambled over uplifted granite, dragged rock sample bags, worked through the almost 24-hour daylight and drew detailed coloured maps. He had a knack for fixing machinery and an ability to improvise repairs when necessary. His cooking skills were less developed. He buried bacon and caribou leftovers, then dug them up later if food supplies ran low. His field team had to brace themselves for his breakfast specialty—canned sardines folded into any surplus oatmeal.

Stu was attached to GSC headquarters on Booth Street in Ottawa, but he appeared there only in the winter, when his field explorations were curtailed by snow and frigid weather. Then he took his few possessions out of storage, rented some accommodation and hid out in the departmental library, poring over aerial maps of geological formations and writing reports. Few people knew him; even fewer ever saw him. He did run into a junior employee who described his summer job with a company that worked for the De Beers Group. The work involved sifting bags of glacial debris looking for diamond-indicator minerals in the area between Hudson Bay and the Great Lakes.

By the late 1960s and early 1970s, the GSC was using helicopters to survey rough terrain. They had been used by the Americans during the Korean War, from 1950–53. Fast and agile, they could skim over the ground, land a geologist to check things out and then fly to another spot. In 1952, the first helicopter-supported geological expedition had set out over the rocky, treeless Barrens. Previous estimates had concluded it would take 400 years to compile a basic geological map of Canada. That year, 147,630 square kilometres (57,000 square miles) of the eastern Barrens were mapped in one flying season, and the estimate for completion was reduced to 25 years.

Stu had found another interest: flying. But helicopters could be tricky in the mountains. Mist, cliffs and downdrafts were dangerous and could cause a pilot to lose control and crash. Still, Stu spent hundreds of hours with US Air Force pilots, who had a contract to fly helicopters in the Rockies. He compared the helicopters to dragonflies, setting down on mountaintops, rising alongside rock cuts, hovering over stream beds and catching thermal updrafts. Up was good; down could be bad.

After 10 years with the GSC, Stu headed a team of 12 scientists who had searched from the Yukon into the Sayunei Range in the Northwest Territories without finding any mining possibilities. He had, however, succeeded in getting his pilot's licence. It didn't do him much immediate good because the government wouldn't authorize an aircraft for him, and he couldn't afford to buy a helicopter, but he did buy a small float plane to fly from one high mountain lake to the next and photograph rock formations.

One day, Stu received a radio call from another team leader working deeper in the mountain wilderness asking if he would take over teaching a student for a while. The caller seemed fed up, so Stu agreed. A few days later, a plane landed at his lakeside camp, and out onto the pontoon stepped a short, muscular teenager attired in a red vest with multiple fat pockets, a camera and compass hung around his neck. Two huge stuffed backpacks followed. He managed to raise both arms over his head, smile and wave like he was meeting his best buddy. He then started walking along the pontoon to the front of the plane, where the propeller was still turning. Camp personnel screamed for him to stop, but he kept on walking and waving. They turned away to avoid seeing the inevitable mess. He missed a step, fell off the pontoon and landed upright, submerged in icy water up to his neck. Chuck Fipke had arrived. He irritated Stu Blusson no end. He never shut up. He was impulsive. He paid no attention to detail. Chuck was given the menial job of colouring the scientists' bedrock maps—granite in red, limestone in blue, shale in brown. He scribbled all over, inside and outside the lines. Blusson blew up, then he gave up.

Chuck found Stu inspiring. Only eight years older than Chuck, Stu was a pilot, could identify ground rocks from the air, had a doctoral degree and could live off the land. He also played the stock market. Chuck learned that Stu had once taken a flyer on a mining stock, made a million dollars, then lost it all when the stock crashed and he did not sell in time. Even though Stu had lost the money, Chuck was so impressed that he had made it in the first place that he automatically scarfed down a whole bowl of Stu's breakfast specialty without thinking or throwing up.

Stu was the one who had banished Chuck to the mountaintop to help the paleontologist collect trilobites. Landing by helicopter, Chuck was enthusiastic. He loved collecting minerals and smashing rocks to free the ancient specimens inside, but he was not allowed to keep any of them. As the pair waited in the tent for the tardy helicopter to pick them up, Chuck talked, quizzed the expert about trilobites and spread their collection all over the tent. The paleontologist drew a line down the centre of the tent floor and told Chuck not to cross it. The only thing they were sharing was the last bit of peanut butter.

Stu was accompanying the pilot to pick up the trilobite duo and had been taking some notes when their helicopter's rotor malfunctioned. They began spinning around, then started falling to earth. The pilot shouted and tried to regain control. They were coming in upside down, but the pilot managed to change the blade pitch and right their craft just before they landed hard in a small ravine. The rear transmission blew through the firewall behind them, whizzed past their heads and smashed out the front of the bubble. They unsnapped their safety harnesses and staggered out.

Smoke wafted from the engine. The pilot swore he was never going to fly again. Stu collected the broken pieces of the pencil he had squeezed in his hand and tucked them away as a keepsake. However, two of his team were still on the mountain, and this helicopter was not flying. The base camp would have to send out another to rescue four people instead of two.

After the ordeal, Stu decided that Chuck did have some positive qualities. He was optimistic—a necessity when working in isolation—and resourceful, and he was still smiling after his mountaintop adventure. The two became friends, although they were polar opposites in personality and lifestyle. Chuck was thickset and myopic, with thinning hair, a boisterous laugh and an irritating habit of forgetting where he had left things, such as his rock hammer or his glasses. He had a predilection for booze and strip bars. Stu was lean and quiet. He was affiliated with government scientists and stockbrokers, and possessed a flying licence that came in handy for searching vast swaths of the Barrens. He had survived some impressive brushes with death in the backcountry and even sported a scar on his left arm from an encounter with a grizzly bear during his early years in the Yukon. Stu had been mapping bedrock one summer, identifying mineral deposits for the GSC. While chipping away at a fossil, he heard an unusual sound and spotted a young bear coming out from behind a boulder and crashing past him. Regaining its balance, the surprised bear lunged toward the even more surprised and startled Stu, who whipped his camera out of his backpack and pumped off some pictures. Without warning, the bear turned and fled. It wasn't until later that Stu noticed a deep claw track on his left arm.

Stu and Chuck shook hands at the end of the summer of 1969, but Chuck would finish his studies and travel around the world before they met again. The pair kept in touch though. Stu was still working in the Mackenzie Mountains and had enough success in the stock market to buy his own helicopter. He was branching out into freelance consulting and prospecting, just as the GSC was pulling back from exploration and focusing on refining maps. It was time for Stu Blusson to move on.

4

Out in the Jungle

IN 1970, CHUCK FIPKE, NEWLY graduated heavy-metal geologist, landed in Australia and was dispatched to Papua, New Guinea, to check out a copper deposit at Ok Tedi in the middle of dense jungle. Kennecott's advanced technology allowed the company to assess the commercial value of deposits at low cost by analyzing for heavy indicator grains containing copper minerals. Chuck was now able to work with this technology and realized it could be put to additional uses, such as searching for gold, nickel, silver and zinc. The other geologists used their time off to get as far away from the jungle as possible. Chuck used it to explore the country with Marlene and Mark in tow; otherwise they stayed in town.

Each work day, the company helicopter picked Chuck up and dropped him at the work site, then flew him to the next site and the next and so on. During days not spent in the field, Chuck worked in a hot, humid shack checking maps and samples. Often the region was so wild that a ground landing was impossible. The helicopter hovered as close to the ground as possible, and Chuck tucked his head down and jumped out. One day, after his ride had left and he was ready to set out on his explorations, he turned around to see what looked like 20 Stone Age warriors with bows strung and arrows pointing at him. Chuck raised his arms over his head, then started taking off his brightly coloured vest. Slowly he handed it to the man he later described as "the one who looked like the chief." One of the warriors gave Chuck a shield. Chuck handed over his gloves and got a feathered necklace in return. The gift exchange continued until Chuck was left standing in his undershorts and boots, but he'd gained what he called "an amazing collection of stuff," including shields, spears, bows and arrows, feathers and talismans. When the helicopter pilot returned, he was stunned to see the almost naked geologist waiting to be picked up along with all his treasures.

When he was 24, Chuck contracted malaria and could no longer do fieldwork. When his condition worsened, the company flew the Fipke family back to Australia, where Marlene nursed him when doctors gave up on his recovery. Day after day, she fed him by hand until finally his eyes opened, he looked up at her and spoke her name. Her prayers had been answered.

Back in good health, Chuck was ready to start exploring again and easily found jobs in the Australian mining industry. Sporting a newly cultivated goatee, he prospected for tungsten and dug opals in the outback after walking in and applying at Broken Hill Proprietary (BHP). But Chuck still had wanderlust, so he took his family and his job skills to South Africa because he wanted to see the world-famous diamond mines at Kimberley. While there, he asked De Beers for a job. After a security check to prove he was not a spy, Chuck was shown around by a Dutch geologist. He talked, laughed, asked a million questions and was given some kimberlite samples for his growing personal mineral collection. He put the material in a nearby glass jar and carried on talking. By the time they returned to the front office, the hiring manager had left, so Chuck found a job with another company exploring for antimony and never did get back for a De Beers interview. While they were in Africa, Chuck decided the family should see some of the continent, so they travelled through Malawi, Namibia and Rhodesia (now Zimbabwe), Uganda and Zaire. They met pygmies in the rainforest and bushmen from the Kalahari.

By 1975, the Fipkes were in Brazil. Chuck got a job exploring for zinc with Cominco (Consolidated Mining and Smelting Company of Canada, now Teck Cominco), which was based in Vancouver. While panning in one of the many rivers running through the jungle backcountry, he saw not only zinc-indicator minerals but also some diamond-indicator minerals. Something was going on here; they looked a lot like the kimberlite samples he had in his glass jar at home in Rio, where he had been conducting his own personal experiments with heavy mineral separation on his balcony. He asked Cominco to let him search for diamonds in Brazil. They refused. He had no skills for such exploration, they were not impressed with his lab work and he had caused a furor on a return flight to Rio when some chemicals in his luggage had leaked.

Cominco did not renew Chuck's two-year contract, but Marlene was pregnant anyway and wanted to go home to British Columbia. Chuck decided to set up his own laboratory there to separate heavy minerals for exploration companies. They took the bus home, north through South and Central America and all the way to Kelowna. In 1977, Marlene took half their $12,000 savings for a downpayment on a small house; Chuck took the other half to open CF Mineral Research Ltd.

Back in Kelowna, the extended family situation had improved. Chuck and his brother Wayne now went out drinking together, their dad had stopped drinking, and Marlene's dad had decided to forgive Chuck for getting Marlene pregnant out of wedlock. Chuck built a sifting machine in the back yard and attached a garden hose to wash the samples. Marlene dried them in the kitchen oven.

Mark, who had only known a nomadic life, spoke fluent Portuguese and knew a lot about geology. After arriving in Kelowna, Mark got a newspaper route, but his father borrowed his receipts to buy lab equipment. Chuck spent 12 hours a day pitching his new business to every industry person he had ever met and to even more people who were strangers. He pitched a former student colleague from UBC who now worked in Arizona for the minerals division of Superior Oil, telling him that CF Minerals could separate out whatever indicator grain they wanted out of large samples. Forget little bags of sediment, he said. Send in tons and the large volume will point the way to new finds. Chuck got some small contracts, mostly for copper, but the income barely covered the $500 Mark owed the *Kelowna Daily Courier.* It would be years before CF Mineral Research Ltd. would become a leading player.

5

The Superior Clue

DESPITE BEING REBUFFED BY COMINCO, Chuck Fipke had not given up on his dream of searching for diamonds. Superior Oil Minerals Division hired him, and they already had a search method. A few years before, a geochemist named John Gurney, working with company funding at the University of Cape Town, South Africa, had hypothesized, like Chuck, that certain common minerals could reliably form alongside diamonds. Gurney used an electron microprobe to analyze mineral samples from kimberlite pipes, where diamonds can sometimes, but not always, be found. He discovered that the presence of chromite, ilmenite and high-chrome, low-calcium (G-10) garnet could foretell the possibility of a prosperous strike. After examining many South African pipes, Gurney published a paper on the subject. Chuck had heard about the work during his South African mine tour and amalgamated it with results from lab work coming out of Russia and his own experience with field sampling.

Superior knew Chuck from his gold-mining work for them, so when they needed someone to explore for kimberlite pipes northwest of Fort Collins, Colorado, Chuck was their man. Although he found some, like the majority of kimberlite pipes, they were not commercially viable.

But Chuck looked deeper. He knew that the pipes, whether or not they contained diamonds, had formed in a craton, a dense old slab of continental plate. Kimberlite pipes were created when magma bubbled up through a craton, expanding and cooling along the way and forming either a carrot-shaped pipe to the surface or a wide, flat underground structure called a dike. He also knew that the craton he had found in Colorado extended north to the Rocky Mountains in northern Canada.

Meanwhile, Stu Blusson's talents as a consulting geologist had been widely recognized by this time, and he explored for gold both for himself and for contract clients. He was now also very popular with the ladies, a situation that initially surprised him even though he was tall, slim and single, had a professional occupation, a doctorate, investments, lived the outdoor life and flew airplanes around North America. Friends tried to fix him up with compatible acquaintances, and a pilot he knew matched him with an attractive Air Canada flight attendant named Marilyn Ballantyne, who shared his love of the outdoors. They were soon married and bought a house near Vancouver. The Blussons visited Kelowna on many occasions and stayed with the Fipkes, their children and the family dog.

Soon after he married Marilyn, Blusson left the GSC to search for deposits of gold, copper and other economic minerals from Mexico to the United States and the Canadian Arctic, piloting fixed-wing aircraft and helicopters over immense tracts of land, lakes and mountain terrain. His explorations led to the staking of mineral claims and the establishment of Puffy Lake Gold Mine, north of Flin Flon, Manitoba, by Pioneer Metals Corporation.

Stu's extensive knowledge of Canadian and international geology had led him to agree that conditions were favourable for diamonds to exist in Canada, but were they present in sufficient quantity to make recovering them commercially profitable? To find out, he developed an exploration plan using scientific methods. By identifying diamond-indicator minerals in the debris left by the retreating glaciers of the last ice age and tracking the glacial debris to its source, the location of diamond-bearing kimberlite pipes could be identified.

In 1979, with Superior's financial backing, Chuck Fipke and Stu Blusson teamed up to hunt for diamonds. There were many players in the diamond search, from leaders to followers, working toward the common goal, but John Gurney and Hugo Dummett gave a lift to Chuck and Stu's quest. Hugo Dummett was a respected authority on porphyry copper deposits and an economic geologist, described as a man of "brains, ideas and energy" in the search for economic diamond deposits in North America. In 1977, Dummett had moved to the United States as senior geologist for Superior Oil/Minerals and began to explore for diamonds. He mentored a new generation of diamond geoscientists and promoted the release of once-proprietary exploration methodologies, like diamond-indicator mineral geochemistry, to improve the chances of locating commercially viable deposits. In addition, as a strong advocate of corporate social responsibility, Dummett helped flow the economic benefits of mining to northern and Aboriginal communities.

Without John Gurney's astute academic observations and accurate interpretations about the association of diamond-indicator minerals and ore-grade diamond deposits, Chuck and Stu might never have put all the jigsaw pieces together. Even with their vast resources, De Beers' exploration teams were unable to solve the puzzle that Gurney had already deciphered: diamond-indicator minerals and their chemical compositions were crucial in the hunt for diamonds. De Beers was on the wrong track until they adopted Gurney's methodologies, and even then they failed to understand the significance of glacial movement direction in northern Canada when looking for diamonds, which hampered their search. Chuck understood, and it gave him a head start.

But there was something else besides knowledge of his home turf in Canada that gave Chuck an advantage. When he got his hands on two previously secret reports that Superior had jealously guarded—a diamond study and a diamond-bearing kimberlite study—they contained the exact information he needed about the elemental composition of indicator minerals associated with diamond-rich kimberlite pipes. Armed with this information, Chuck knew it was time to go back into the field.

In 1981, Superior Oil/Minerals still had an ongoing interest in diamonds, although the company's primary focus was now the search for gold. Backed by Superior, diamond prospectors Chuck and Stu were doing their endless summer fieldwork, this time in the Sayunei Range near where they had met many years before. When a college friend of Chuck's mentioned a strange pipelike formation in a remote location, Chuck was interested. Chuck and Stu decided to stake a claim there, calling it their Mountain Diatreme, from the Greek word for a volcanic vent or hole in rock. Stu recalled how he had taken aerial photos there years before for the GSC but never noticed the pipe. Now he retrieved a photo that showed another round spot on a ridge. Maybe there was something there.

When the regular supply plane arrived, the pilot chatted with Stu and Chuck about a big camp set up over on Blackwater Lake. The company working there seemed to have plenty of money to buy flying time and expensive supplies. After some casual conversation and questions, they deduced it must be De Beers looking for kimberlite pipes. Stu and Chuck decided this called for a reconnaissance mission. That night, the two of them flew over to the Blackwater camp in Stu's helicopter, dropped down and skimmed through the valleys to deaden their noise. They flew low over the babbling river near the camp at about 2:00 AM and landed on a sand spit. It was quite light out, but this was like a commando action behind enemy lines, and they got a shot of adrenaline from the thought.

Chuck jumped out and dug a sample from the spit, while Stu watched the surrounding rocks, keeping the engine running. They moved forward and got another sample, then flew off to an area they thought was outside the De Beers claim boundary, cut the engine and listened for any sound of discovery. All clear. They dug more samples. Stu noticed the large amount of glacial debris that had been chewed up by ice and strewn around the area. After a couple of hours' work collecting samples, they flew out to Norman Wells, where they had breakfast and put the samples on a flight to Kelowna and CF Minerals. When Chuck got home, he tested the samples himself and found they were loaded with chrome diopsides, garnets and black ilmenites—in other words, diamond-indicator minerals. Chuck called Superior Oil/Minerals and chatted with Hugo. Meanwhile, Stu contacted his old GSC colleagues who had just mapped the Blackwater location. From his initial observations, Stu believed De Beers had not staked the right place. The ice had moved from south to north, and Stu thought they should stake the ground some distance away from the De Beers claim. In October, after Diapros, a De Beers affiliate, had left the area for the season, Chuck, along with five samplers and a pilot, got to work along the Blackwater River, preparing to stake their claim. Stu was away on a consulting job. The claim was filed in Yellowknife under the name of Marlene's father, Harley E. Pyett, who had agreed to act as cover.

Then Hugo Dummett called Chuck with bad news: Superior was pulling out, saying there were no diamonds under the Rocky Mountains or the northern tundra. The company had spent $11 million and found nothing, so made the decision to leave Canada and focus on the United States. Without joint venture money from Superior, Chuck had to rearrange financing once again. But he did ask Superior if he could get the Blackwater claim—after all, if they needed a partner again to fund a discovery, Hugo would be on top of the list. Superior agreed to relinquish the claim, and CF Minerals got title to the mountain properties. The Blackwater claim went to Chuck and Stu jointly as the Blackwater Group. They would be equal partners forever.

Chuck and Stu still needed to earn a living, so during the winter and following summer they undertook individual consulting contracts, sometimes working together and sometimes not. Once, flying together on a packed commercial milk-run flight from a zinc prospect, they found themselves assigned to seats in different rows. Near the rear of the plane, Stu was catching up on some reports when he heard a man with a South African accent talk about diamond prospecting. It sounded like Hugo Dummett, but it wasn't. Stu adjusted his seat and pretended to doze as he took mental notes. The speaker was a De Beers geologist. He was telling his Canadian colleague all about exploration along the African coast, in the Arctic and on Greenland. Then the Canadian asked about Blackwater—one of Stu's favourite topics. He got an earful. It seemed De Beers had also concluded that the indicator minerals near their camp had to have been transported up from the south by ice, but that their source was still within the De Beers claim. They had drilled four holes, two of which were promising. They should soon have results. Stu noted the indicator minerals found and was pleased to hear no mention of neighbouring claims. When the plane made its first stop, Chuck's seatmate got off, and Chuck shouted back to Stu to come up and join him. Stu, feigning sleepiness, declined. But there was no more intelligence to be had. The De Beers geologists fell asleep until their final landing in Calgary. Stu couldn't wait to tell Chuck what he had overheard. The find was theirs.

In August 1982, Chuck worked at the lab in Kelowna. Stu and a sampler flew to the west side of Blackwater Lake with his helicopter and camped there. They focused on a nearby valley surrounded by 400-million-year-old dolomite rock. Updrafts and crosswinds made flying tricky. In bad weather they were grounded, so Stu took the opportunity to get in a little fishing. One day when Stu was cutting across a stream, stepping from stone to stone to keep his feet dry, he noticed some large pinkish rocks. Others might have thought them just to be pretty stones, but they were very significant to a geologist searching for diamonds. They weren't local bedrock. They weren't kimberlite either. They were sharp chunks of pink granite. Stu's trained eye knew they were billions of years old. The same kind of granite boulders had been found by the GSC dumped high in the Mackenzie Mountains, and they were geological anomalies there, too. They were glacial erratics torn by ice action from granitic bedrock in an Archean region (2.5 to 4 billion years old) called the Slave craton, shield or province, which hung like an ancient breastplate across the boundary between the Northwest Territories and Nunavut, one of the oldest parts of the North American continent. But new GSC maps showed these exposed rocks did not start for another 323 kilometres (200 miles) to the east. Suppose they had been carried west by glacial movement in northern Canada? If the granite boulders could be transported this far, so could kimberlite and its diamond-indicator minerals. That meant they were not at the origin of the glacial dispersal train but what geologists call "downstream" from it. They were looking in the wrong place. Finding the indicator-minerals dispersal train's point of origin in the east would unlock the mystery of where diamonds could be found in the Arctic. The hunt was on.

Glacial terrain north of Lac de Gras, Northwest Territories.
LYNDA DREDGE. NATURAL RESOURCES CANADA, EARTH SCIENCES SECTOR 2001-184

That serendipitous pink-granite discovery started an 800-kilometre (497-mile) pursuit of indicator-mineral trains back along the path of the glaciers across the tundra. Stu left the sampler at their claim site and took to the air. He flew "up ice" along what he supposed was the glacial route and saw rows of directional pink boulders, as well as grey schist, which he knew also came from the Slave craton. If they were all looking in the wrong place, the Blackwater Group had better figure it out first. He touched down and took mineral samples. They broke camp and flew to Palmer Lake, where Stu left his helicopter, having arranged for another pilot to pick them up and drop them off in Fort Simpson. That would avoid any informers in Norman Wells passing on information to the competition. Back at the lab in Kelowna, Chuck rushed the "B" (code for Blackwater) samples through analysis. They were full of diamond-indicator minerals.

In September, Stu went back to Palmer Lake, started up his helicopter and flew alone for days, heading east away from the mountains and over an endless expanse of low, swampy ground. On the last day, he could have lost it all. It was getting dark, there were no ground references on the endless bog, he had just enough fuel to get directly to Fort Simpson, and he could not afford to make a navigation error. Setting down among the dead trees and hummocks would flip his helicopter and probably kill him. If he did manage to crashland and survive, nobody would know where to look for him. Finally, one tiny light appeared below him, then more. Stu and his 20 collected samples landed at Fort Simpson. There, to his surprise, he ran into Chuck, who told him he looked like hell. Chuck had come up to stake more ground. "But we don't need more ground," Stu snapped. "You need to look at the samples first." Chuck, however, thought they had agreed to get more ground to the south and east. He was so fixated on that idea that he stayed a week at the trailer motel in Fort Simpson, watching TV while waiting for the weather to clear so his chartered helicopter could take off. Stu was long gone, and once again the money situation was tight.

When he returned to the lab, Chuck processed Stu's

samples. Every sample had diamond-indicator minerals, but there was no discernable pattern or source. It was not a good sign; however, they had kept in touch with Hugo Dummett and he offered to have the indicator minerals microtested in a skunk works project he had in the United States. (Skunk works projects are run by small groups of experts who work outside of the mainstream to develop experimental technologies rapidly or in secret.) The chemical analyses would yield detailed information about what they had found and what it might mean, and the results could be kept quiet. It was a big saving for Chuck and Stu because testing one gram of indicator minerals cost $1,000. The results were good: many of the garnets were G10s, indicative of the presence of diamonds. Now Chuck and Stu had to decide whether to keep looking for diamonds at Blackwater or follow the glacial trail. The ice had advanced and retreated many times during glacial freezing and melting, but the principal direction was from the east, so that was where they headed. The Blackwater claim lapsed.

Chuck and Stu's next plan was simple but time consuming. They would collect rock samples from glacial features, such as eskers, during the summer season and analyze them for key indicator minerals back in Kelowna during the winter months when the Barrens were frozen.

After coaxing another $30,000 from investors, the two geologists hired a helicopter and set out. Stu would not be the pilot. It would be faster if they worked as a sampling pair. Just to make sure nobody followed to see what they were doing, they met the pilot outside Whitehorse and didn't tell him the route until they were airborne. They returned to the bog Stu had flown over, probably the remains of a meltwater lake that had been formed by receding ice, then flew on to Lac la Martre, the farthest point of this sampling tour. About 30 kilometres (19 miles) past the lake, they saw low humps of pink granite protruding out of the tundra, just like the stones they had seen in the streams around Blackwater Lake. At this location they were closer to Yellowknife than Norman Wells, so they asked their pilot to put them down there. The pilot had never once asked them what they were looking for, but in making his goodbyes he mentioned he had another job taking some geologists up where they had been. Chuck and Stu were stunned. Was someone on to them? One look at the pilot's logbook would reveal their route.

Much to the chagrin of its geologists, De Beers management in South Africa had discounted preliminary findings around Blackwater Lake and chosen to focus their diamond quest farther south in Alberta, Saskatchewan and around Hudson Bay. But Superior Oil/Minerals once had an agreement with Falconbridge Limited, a base metals company. Were they sniffing around for clues? And any number of junior mining companies could be slipstreaming the exploratory trail of Chuck and Stu.

Then there was Clifford's Rule. In 1966, T.S. Clifford wrote that diamonds and kimberlites were to be found in

Archean rocks over 2.5 billion years old. These rocks form the tectonically stable cores of continents known as cratons, and diamond exploration should focus on those areas. However, there are exceptions. In Gambia, diamonds are found with early Proterozoic rocks (2.5 to 1.6 billion years old), and in eastern British Columbia, diamonds are found in rocks that may be underlain by fragmented Archean (also called Archeozoic) rocks (3.8 to 2.5 billion years old) that are parts of shattered continents reformed.

Over time there are fewer and fewer areas of possible diamond potential that remain unexplored, so diamond prospectors need to look in places that at first glance appear not to follow Clifford's Rule. For decades, nobody thought of looking under the tundra because it didn't seem to fit the rule. Chuck and Stu believe that diamond occurrences were probable in the Northwest Territories, followed Cliffords' Rule and, for that reason, initiated reconnaissance indicator-mineral surveys. Even though first results were confusing, they persisted when others gave up and moved on. It was this tenacity combined with their skilful application of Clifford's Rule that worked for them.

Back at the lab, work was piling up. Chuck had to process clients' samples as well as his own. There wasn't much in the bog samples from Glacial Lake McConnell, but past Lac la Martre things got interesting. There were a lot of garnets, and Dummett's microprobe showed they were G10s. However, the investors were not impressed and refused to pony up any more cash. Chuck and Stu did not have enough money to fund the project themselves, so Chuck worked at the lab, and Stu took on consulting assignments. Their great diamond quest was stalled.

Then Hugo Dummett came up with a very upsetting premise. Suppose the pipes once existed but had been chewed up by ice and were now wiped out. The very thought drove Chuck nuts. Stu was more philosophical, suggesting that if the pipes were really gone, they could author an amazing scientific paper proving they had once existed. Chuck thought Stu had lost interest in the venture—maybe he didn't want to risk his financial security on the search or endanger his life out on the tundra. Stu thought Chuck had lost his perspective on a chase that might go nowhere and that he was too attached to an elusive monetary outcome.

Chuck was stressed. His lab employed 36 employees at peak times, but it was boom or bust. Sometimes there were long periods when the work dried up and he had to lay off staff. Chuck saw his own life replayed before him when his oldest son, Mark, quit school and married his pregnant girlfriend. Marlene complained that he was never at home for their four other children, and there were many things that needed repairs around the house. Instead of replacing a light socket, he bought candles. Wrinkles and a receding hairline made him look older than his 36 years.

But then Chuck hit upon a new idea. The original Mountain Diatreme experience in the Mackenzie Mountains had been a bust for diamonds, but maybe other metals like gold, copper or zinc could be mined there. Chuck would form a company to re-explore the area. He called his creation Dia Met, for diamonds and metals. Chuck's brother Wayne, whom Chuck had beat up on many occasions in their youth, was still enthralled by his brother's adventures and convinced a wealthy acquaintance to invest in Chuck's start-up company. Wayne bought $2,500 worth of shares, while his well-heeled friend, Dave Mackenzie, invested $50,000. Chuck got the most shares by signing over his claims to Dia Met, and Stu received a block of shares and a directorship in recognition of his role in finding the mountain claim areas.

Dia Met Minerals Ltd. went public on the Vancouver Stock Exchange (VSE) in 1984, but it was just one of hundreds of penny mining stocks, most of which would crash and burn. No stockbroker would touch Dia Met, especially after Chuck paid them a promotional visit arrayed in his one-and-only creased blue suit and crooked glasses, giving them a spiel about diamonds. After many rejections, a broker who knew Stu agreed to sponsor the offering. Shares went for 50 cents and some venture capital funds, and wealthy west-coasters bought in, mostly because Chuck was a real prospector and there were tax benefits: for every dollar they lost on a high-risk mining stock they could write off $1.33. They lost. The stock dropped to nine cents a share.

Still more money was needed to keep the company solvent. Chuck gathered his friends who had backed the

CF Minerals start-up and proposed new partnerships. They anted up and threw more money into the pot. Chuck and Stu carried on lurching from financial crisis to financial crisis.

Then things got worse. Superior Oil, including its minerals division, was sold in its entirety to Mobil Oil. The diamond program was killed, Hugo lost his job, and Chuck and Stu lost their free microprobe analyses. By 1984, they had sufficient finances for limited heavy-minerals sampling to try to determine the source of the diamond-indicator dispersal train. By 1985, they believed they had it nailed down. But they weren't there yet. It took until mid-1988 to process all of the samples and plot the pattern.

Chuck and Stu assessed their situation. They agreed they had only one more kick at the can. Just 160 kilometres (100 miles) beyond their last sample site in the mountains lay the Barrens. They had to go there immediately and track the path of the ice with its diamond-indicator train right back to the point of origin. They had to do it fast, and they had to pay for it themselves.

CHAPTER

6

Bingo!

CHUCK AND STU HAD NEVER worked in the Barrens before, but how difficult could it be? They had worked in every other conceivable terrain on the planet. The tundra was flat—easy in, easy out—and you could see for miles. Over the winter of 1988–89, Stu paid a return visit to the photo library at the GSC in Ottawa and scanned hundreds of black-and-white topographical maps of the Barrens prepared over many decades. These maps were based on photographic images from aerial surveys. And one feature stood out: the great ridges of eskers, like those where they had previously found indicator minerals. Contact sheets highlighted the white channels, the grey swamps and the black lakes. They had only to follow the ice sheet. They discussed where to start.

Stu wanted to begin south of a big, irregularly shaped body of water called Lac de Gras and move east. To save time and money, they would rent a float plane out of Yellowknife, and Stu would fly. They would set down on lakes just as they had done in the past and cover a lot of territory.

Back when Chuck had started Dia Met, Stu had established his own public company called Pioneer Metals of which he was chair and chief geologist. His investors had put in money for a possible gold discovery in Manitoba, and in the spring of 1989, Stu was on site supervising the project. He was unable to make the July 1 start date for the Barrens exploration. Mid-July passed, then August 1, and Stu was still stuck in Manitoba. After mid-August, Chuck decided to go it alone and hired a Yellowknife air charter company, saying he was looking for gold. Flight plans were approximate; Chuck navigated, and all the pilot had to do was fly. Chuck talked vaguely about finding gold, paid in cash from a roll of $100 bills and got receipts with the "received from" line left blank. Stu's plan to sample along the eskers was replaced with Chuck's plan to fly over the tundra in north-to-south lines. He worked from the GSC Glacial Map of Canada, which showed only general ice-flow features.

The Cessna Chuck had hired carried extra fuel, and they returned to Yellowknife every night. Every morning, he told his pilot the heading for the day. Whenever he saw a promising site where landing was possible, the plane set down on the water and drifted to shore. Chuck hopped off the pontoon in hip waders and dug samples from eskers, lakeshores, riverbeds and rocky outcrops. The pilot stayed inside the plane, away from the attacking insects. Chuck scooped up a shovelful of gravel, pitched out the big rocks he did not want, slopped the rest into a bucket with water and dishwashing detergent, gave the samples a good wash and put them in his bag. Then they flew on to the next area that caught his eye. Most diamond prospectors threw away the sand and kept the rocks. Chuck, who became known as "The Sandman," did the opposite. But he didn't care what they called him. Somewhere here would be diamonds. The plane was refuelled in Baker Lake or Yellowknife, and Chuck put the samples in storage but never stayed to chat. Prospectors were known to be a secretive bunch, but Chuck Fipke, or whatever name he was using, was exceptionally secretive and suspicious.

Many places where Chuck sampled had no name, so he wrote in the map coordinates and any nearby features. One day near the end of his sampling marathon, he instructed his pilot to fly over Lac de Gras, the source of the Coppermine River. There was no sampling site on the rocky edge or on the north boundary, but then Chuck saw a big, S-curving, east-west esker protecting a small lake with a sandy shore and a dark striation in the sand. Interesting stuff. They landed and drifted to shore. Chuck dug up some dark-purple sand and bagged it. Then he decided to enjoy the view and climbed the side of the esker to its top, broad as a runway, and looked out over kilometres of Arctic tundra etched with caribou trails and falling away to the horizon. The breeze up there blew the bugs away or at least set them off course. After taking in the view, he ran back down and jumped into the plane with his last sample bag: G-71.

Over the decade, in one manner or another, Chuck had worked his way across the Arctic tundra, and now they were off to the Glacial Divide just west of Baker Lake. East, ice flowed to Hudson Bay, west it flowed to the northern islands, south into central Canada and west into the Mackenzie River and Blackwater Lake, where he and Stu had discovered De Beers exploring for diamonds. Using an assumed name and, when necessary, a fake company called Norm's Manufacturing, in honour of a CF Minerals employee who had never been north, Chuck chartered a variety of float planes and used many different pilots so nobody would put two and two together and discover what he was up to. He flew repeatedly over the Arctic Circle, Coronation Gulf, Lac la Martre and Great Slave Lake. From August 17 to 27, he dug 96 sampling holes.

The broad survey and sample collection was now complete from Blackwater Lake to the Glacial Divide, an immense distance of 1,200 kilometres (745 miles) and an area of 520,000 square kilometres (200,720 square miles). Somewhere out there was the diamond cache. Now it was all in the lab work.

Bingo!

An aerial view of an esker and till plain north of Lac de Gras, Northwest Territories.

Chuck left for Kelowna on August 28. Back in his lab, he used the chemical testing procedures he had developed. Even though he kept his lab technicians busy, he did the really crucial tests himself. One could never be too careful when the subject was diamonds. In his locked office at CF Minerals, he plotted the locations of the indicator-mineral concentrations the tests had identified on maps, and he tried to track them back to their origin. But that part was an inexact science. There were gaps in the trail, and he had to make educated guesses about which way the glaciers had dragged the indicator-minerals. He had persevered through all the investor fatigue, setbacks and false starts, doggedly plotting the trail, keeping silent and moving forward.

Stu was stunned that Chuck had gone ahead without him and became even more agitated when he found out the random manner in which he had sampled. Except for the esker, it appeared haphazard and costly. How much had all that flying cost? Chuck admitted he had gone to Wayne Fipke's friend Dave Mackenzie, who had given him an additional $67,000 for a share in any discovery. They had a new partner. Stu was incensed. Chuck's action had diluted the shares of all the investors, themselves included. But there was more. In order for Mackenzie to realize certain tax advantages and have a tangible asset, he had converted his investment into Dia Met stock. So Dia Met, a public company, was now part of their private venture. "Give the money back," said Stu. "Not possible," said Chuck, "the papers have been filed." Relations between Stu and Chuck were as icy as the glaciers they had been tracking. Stu resigned from the board of Dia Met, Mackenzie became president and Chuck was chairman. When there was a lull in the fighting, both Chuck and Stu knew they needed to find a senior partner with deep financial pockets, but there were no takers. They were out of money, anxious about unknown competitors and distrustful of each other.

Chuck refocused on lab work. Peering into the concentrates and noting his findings on the maps, he saw the pieces coming together. The bigger picture began to resolve into precise details. The trail of indicator minerals, often so small that only his patented concentration techniques could have spotted them, began at Blackwater Lake and spread east. Then, 320 kilometres (200 miles) northeast of Yellowknife, near Lac de Gras, it stopped. Bingo! Chuck could actually recall looking down at the area from the air. Slightly north of Lac de Gras was a shallow lake called Exeter that was bounded by that high esker separating it from another lake, which had no name. He could see the strip of gravel beach running between the ridge to the sheltered shoreline and recalled sampling at the foot of the esker near the beach before climbing up for a look around the surrounding Barrens. That sample contained some 1,500 chrome diopsides and 6,000 pyrope garnets. Two other samples from nearby confirmed the findings. He had found his field of pipes.

But now Chuck needed a better lab and more equipment to determine where the pipes were located before he could start staking. Even more important, were they even worth staking? Were there enough diamonds to justify all the upfront expense? Chuck had very limited resources and could not afford to be wrong. After all, he had experience with diamondiferous pipes that had proven uneconomical to mine, like those in Colorado, and volcanic pipes that contained no diamonds at all, like the Mountain Diatreme, the 600-metre (1,970-foot) kimberlite pipe in the Mackenzie Mountains southwest of Norman Wells.

What Chuck needed was a scanning electron micro-scope to analyze the indicator minerals. Luck was with him, but it came with strings attached. The federal government had a new program called the Scientific Research Tax Credit that would allow Chuck to buy the microscope at no cost after tax. Dia Met investors agreed, and Chuck and a colleague went off to New Jersey for a quick course in how to use it. All winter he worked in secrecy performing grain analyses. By spring, he knew there were at least three hot "smoking" pipes that had all the superior attributes of the finest diamond mines in the world. Chuck knew that pipes usually come in clusters of 25 to 35, of which only two or three would be rich enough to mine. At an estimated value of $70 a ton of kimberlite recovered and assuming 400 million tons, the pipes spec'd out at about $28 billion. But he had no money to locate the pipes and stake them. Government tax-shelter rules stated investor money had to go into existing property, not exploration. With no money to stake claims, he was ineligible for financing. The find was open. De Beers or any number of mining companies could scoop it up in an instant. Meanwhile, Chuck needed to keep doing analyses for his oil- and mining-company clients.

But who could he trust? On a trip to New Zealand to consult on locating platinum, Chuck had met a man called Dawson, who was an explorer, hunter and beachcomber, and invited him back to Canada. Dawson showed up and crashed on the Fipkes' sofa. Chuck decided that he, Mark and his beach buddy Dawson would go back up north for more sampling. Chuck told Mackenzie about the esker findings, and he offered to fly them there in his Piper aircraft. He had never flown over the tundra before, much less landed there, but Chuck assured him the esker was wide and smooth. They landed in Yellowknife, where Chuck decided to lease a float plane with a pilot, and they flew on to a lake 50 kilometres (31 miles) from his G-10–garnet esker and beach. There they set up a mock claim to throw off any spies. Mackenzie and Dawson did more sampling using the rented plane and pilot to cover an area that spread out from their pretend claim site. Chuck plotted their sampling template to make sure every possible area that was covered and, just to deceive any curious busybodies, he included some improbable areas as well. They took hundreds of samples during the eight-week trip and celebrated their return to civilization in the bars of Yellowknife.

Having had enough adventure in northern Canada, Dawson returned to New Zealand and Chuck to the lab. The amount of information collected was overwhelming, but the only way to make sense of the mix of minerals was if the indicator-mineral train stopped east of Lac de Gras and the indicator minerals were concentrated at the esker beach. The pipe or pipes could be the largest ever recorded, but where exactly were they? Should he start staking now or did he have the right ground? He would need to claim thousands of square kilometres to handle all possibilities, but this was government land. They would have to pay to register the claim, then spend an annual amount prospecting on it. Money was always the problem.

Enter Wayne Fipke. Wayne had saved some money, and his friend Dave Mackenzie was tired of being president of Dia Met. If Wayne became president, he would invest in Dia Met and also convince others to do so. Chuck gave him the job. Stu was sitting on the sidelines because of some legal entanglements, but he thought Chuck should have sampled farther up-ice. Finally, Chuck decided to make camp at the esker and stake out from there.

CHAPTER

7

The Ultimate Find

WITH MONEY TOSSED INTO THE pot from directors, staff, friends, family and sales of new shares in CF Mineral Research and Dia Met, there was enough to finance exploration during the summer of 1989. Chuck and his team of 10 people—friends, relatives and faithful samplers he knew personally—drove up to Yellowknife in three trucks for what Chuck said was a gold-staking expedition. They were ferried to Exeter Lake by plane and pitched their tents on the shore, christening their summer residence "Norm's Camp." Chuck drove in the first stake, then they moved out in all four directions. Keeping the find secret, he'd had the stakes made in Kelowna and flown in. A local order for thousands of claim stakes delivered to the same airport used by other prospectors would have raised unwanted questions. Chuck filed the claims under the name Shirley-Anne Eccott, the wife of a new Dia Met director. All involved in the staking activity were trusted friends and family—but not too trusted. Many still thought they were staking a gold prospect. Even Chuck's lab workers did not realize the samples they had been testing were for a diamond mine.

Chuck was driven by the need to get the claim staked quickly. He still feared his secret treasure would be discovered by others, even though there were no helicopters or planes overhead and, in all his years on the Barrens, he had never encountered Native hunters. This area was too open for the Yellowknives and Dogrib of the Dene Nation, who preferred the forests farther west, and too far from the Arctic coast for the Copper and Caribou Inuit. Still, there would be no breaks in Yellowknife for the crew—no carousing in the strip clubs, getting drunk and blabbing about what they were doing and where. When they were finished, they would fly directly back to Kelowna.

At base camp Chuck had his own tent, which was control and communications central. Each crew member was outfitted with a bug suit, insect repellent, bear spray and a radio for his daily staking assignment. Many of them had never been in the North before. It was a warm, moist summer, and the insects made every day unpleasant. There were three plagues: mosquitoes, blackflies and deer flies. Nothing really worked against the insects, so they resorted to wearing wraparound sunglasses and heavy wool shirts and pants under their bug suits, with leg and wrist cuffs closed with duct tape. In the summer heat and humidity, they were wet with sweat. When they returned at night, Chuck debriefed them. They followed all the government rules. At Blackwater Lake, the rules allowed staking of huge concession blocks, but here they had to stick to smaller claim blocks, although they were webbing together a massive territory with claim markers. Chuck drove them hard, but by the time the money ran out and the summer was over, they had staked only half of the area he thought he needed.

Then Chuck decided he required a sample from the west side of Lac de Gras, near the headwaters of the Coppermine, and sent Mackenzie to fly in and get one. Mackenzie did not want to go. The previous year he had almost ditched his plane when a wheel struck a deep rut. Chuck lost his temper. Mackenzie was shirking his duty. With his Dia Met investment at stake, Mackenzie took off. It was a mistake. He flew the 40 kilometres (25 miles) to the sample site and looked for a place to land. He spotted a short strip of esker that looked suitable and approached, but was hit by a crosswind, then a tailwind. He couldn't land, and he couldn't abort the landing either. The nose caught the ground, and the plane flipped.

Back at his tent headquarters, Chuck was wondering what was taking so long. The float-plane pilot spotted the crash, but there was not enough water to land. He radioed in, and the camp prepared to mount a rescue; however, nobody had a first-aid kit. Then a lone figure slowly walked into sight. It was Mackenzie. He had a gash on his head but otherwise seemed fine. Before the plane flipped, he had shut off the gas and electricals and then found himself suspended upside down by his seat harness. After extricating himself from the plane, he had pulled out a map and compass and made a straight line back to Norm's Camp through water and over hills. Everyone patted him on the back just to assure themselves he was real.

Then Chuck asked the obvious question: "Do you have the sample?"

Mackenzie, who just had survived a plane crash and a hike through bug-infested tundra, thought he was kidding. "No," he replied.

Chuck was beside himself, "You go to all that trouble and you come back without the f-ing sample? Useless!" The next day, Mackenzie was walking the tundra on staking duty.

In late 1989, Chuck made a trip to Vancouver to talk with more investment-fund managers and managed to swing a cash-for-shares trade. He and Mark returned north to work in the snow. They rented a helicopter and hammered stakes into hard ground. The helicopter engine froze, but they now had a claim block of 160,000 hectares (395,370 acres). It was still not enough. Chuck worried that the southern section was not well marked and that other competitors would rush in and outflank them. They were probably safe for the winter, but come spring they would need to move fast.

Chuck, Mark and Mackenzie went back to their Exeter Lake camp in April 1990, but snow still covered the frozen ground. It was difficult to navigate by natural markers like lakes and rivers, but Chuck had a system. He marked out a grid pattern on maps, and from the passenger seat of their $1,000-per-hour chartered helicopter told the pilot, who believed he was working for Norm's Manufacturing and staking for gold, where to go. They flew one claim's width outside their last perimeter and at top speed to each target. Chuck yelled out the stake number. From the back seat, Mark handed the pre-numbered stake forward. The pilot touched down briefly. Chuck leaned out and slammed the stake into the snow. When it held, he shouted, "Go," and the pilot skimmed over the ground to the next coordinate at top speed, often before Chuck could get all the way back inside the bubble. It was a fast way to stake, but it ate up money at a furious rate.

Then something caught Chuck's eye. It was a small lake on Pointe de Misère peninsula, which jutted into Lac de Gras and was the only land feature on his map with a name. There was a ridge on its northern edge, a central depression and another ridge on the south side, and it was within an area they had claimed. Intriguing. What had caused those formations? He asked the pilot to do a quick flyover. They continued staking, and that night at camp, after devouring the meal Mackenzie had prepared, Chuck checked his maps and geochemistry results. He checked ice direction and noted that previous samples taken in the area were down-ice. The lake looked like it had been punched into rock cliffs. It could be directly over a pipe. Should he continue staking or investigate? He needed to prove he had a real kimberlite pipe if he wanted a large mining company to back him. He decided to forget the staking. This was more important.

The next morning, he told the pilot to land on the south shore of the small lake. In below-zero temperatures and a biting wind, they dug through the snow and started chipping at the ice, searching for gravel containing indicator minerals. Hours later they hit flat rock at the bottom. Chuck looked out over the lake. Since there were no indicator minerals or kimberlite here, then maybe he'd find them on the other side of the lake. The helicopter landed them near a windswept spot and Chuck studied the ground. Whatever the glaciers had scooped out of the lake would have been deposited here. The snow was deep, and they had only a collapsible shovel to work with. Mark, Mackenzie and Chuck took turns digging down to the water's edge. With his rock hammer, Chuck chipped the ice, hoping to hit gravel underneath. Nothing. A total of five hours later, they had dug a series of pits, hitting only icy water and flat boulders.

Mark looked up and pointed out a dark area with no snow on the esker. Chuck had discounted his suggestion to dig there before, but now it was a demand: dig there before we all freeze to death. Totally exposed on the high ground, they chipped away at glacial till, sharp chips flying in all directions. Slowly they filled three sample bags, then Mark spotted a pea-sized green stone. He showed it to his father. It was a perfectly formed chrome diopside, a diamond-indicator mineral. Chuck could hardly contain his joy, but he knew he had to remain calm. The pilot was watching. Most likely it had been scraped from the pipe under the lake, and when the softer kimberlite eroded the stone remained. This was his proof—a real pipe. He dropped it into a sample bag, and the trio returned to the helicopter for the ride to Yellowknife.

The unnamed lake was in the centre of Pointe de Misère peninsula, so Chuck called his treasure pool Point Lake. The humour of it all struck him, too. There was another, larger Point Lake northwest of their claims area, so the confusion would work in his favour. Sampling finished, they flew directly back to Yellowknife, then on to Kelowna and Chuck's lab. In every lakeside sample he found a profusion of diamond-indicator minerals and chips of kimberlite mixed with the till. In kimberlite pipes with gem-quality stones in commercial quantities, a concentration of 1 carat (0.2 grams) per 100 tons is profitable; based on these latest samples, Chuck estimated the diamond concentration at Lac de Gras at more than 60 carats (12.0 grams) per 100 tons, with a quarter of those good quality or better. John Gurney examined samples sent to him by Chuck and stated in his report, "The dataset represents the best for diamond potential that we have seen anywhere in the world."

Now Chuck was ready to negotiate with a major

company. Stu was back, after wrapping up the buyout of his partners in other projects in which he had been involved. They approached BHP and Crystal Mines in Australia and Placer Dome Inc. in Vancouver. The more people who were contacted, the more it endangered their diamond secret. But they could not wait forever—they needed money to spend on the staked claims or, under mining law, they would lose them. Finally, in August 1990, Hugo Dummett, who had landed a job at BHP, agreed to talk with them. Dummett had risen through the ranks of BHP to become vice-president of minerals discovery, and Chuck looked to Hugo to convince executives at BHP to back his Northwest Territories project.

Hugo wanted BHP in, especially after he read Gurney's report, but the partners played good cop/bad cop with him. Stu would agree to a clause but then say Chuck would object. Chuck would agree but say he would have to check with Stu. They upped the stakes and wore Hugo down in weeks of negotiations. Chuck and Stu were working well together again, but privately they argued about Chuck's secretive ploys, Stu's many absences, Dia Met's legal role and how the profits from any future mine would be divided. Finally, the deal was done. Chuck and Stu each got 10 percent personally, Dia Met shareholders got 29 percent (which raised Chuck and Stu's amount because they were major shareholders) and BHP got controlling interest at 51 percent for fronting $500 million for exploration and development. Staking would continue at BHP's expense in a 6.5-kilometre (4-mile) buffer zone around existing claims. Chuck remained in control of field operations, and Stu got an option to explore in the vicinity.

Chuck and Stu had been actively exploring for diamonds for eight years and had failed to find a single gem-quality or industrial-grade diamond. Superior had pulled out of the diamond exploration business. Promising pipes had disappointed. De Beers had applied their own indicator-mineral formulae and come up empty. But Chuck and Stu had guessed the solution was simple: diamond-indicator minerals had been transported by ancient glaciers to points far away from the kimberlite pipes. They needed to look "upstream" from where the indicator-mineral deposits had been found to locate where they originated. They had also flown thousands of air miles and hundreds of hours back and forth over the Arctic Circle, and Chuck had used a magnetometer to track magnetic field variations that could indicate kimberlite, before serendipitously finding the hot site they called Point Lake, northeast of Yellowknife and just south of the Arctic Circle.

Now BHP and Dia Met had to deliver the goods. They had expanded the claim block to 388,498 hectares (960,000 acres) but had to come up with a proven pipe. Executives at BHP wanted to know what they were getting for the money they had put up. Hugo told Chuck he had stopped staking and had ordered a drill flown in to their Point Lake. They had to find diamondiferous kimberlite or it was all over.

Scientists work on a till plain north of Lac de Gras, Northwest Territories. LYNDA DREDGE. NATURAL RESOURCES CANADA, EARTH SCIENCES SECTOR 2001-189

Chuck was too nervous to stick around, so he assigned a junior Dia Met board member to stay at Norm's Camp near Exeter Lake off the northwest corner of Lac de Gras during the drilling at code-named "Point Lake." Rig components were delivered by Twin Otter, and a crew arrived from Winnipeg to assemble and operate it 48 metres (157 feet) from shore and directed at a 45-degree angle to the centre of the lake. Again, they were told the search was for gold. With great noise, a diamond-tipped hollow pipe was driven into the lakebed and core after rocky core pulled up and placed in parallel rows in wooden boxes. This continued 24 hours a day for four days. Then the drill foreman came to Norm's Camp with news that they had struck much different, softer rock—kimberlite—at a depth of 138 metres (452 feet). Cores of kimberlite were already in the boxes.

Following Hugo's instructions, they coded any radio-telephone communications. Chuck got a call from camp. "So, how's the fishing?" he asked.

"We found the biggest fish you ever saw in your life," came the reply.

"Well, it's about time," Chuck yelled.

A few days later, a float plane landed with Chuck, Mackenzie, Hugo, Hugo's boss and John Gurney, who had been at a conference in Saskatchewan. Inside a wooden shack, the cores were laid out like bodies and inspected closely through magnifying glasses. Gurney pronounced them "juicy" with dark garnets and other minerals. There were no diamonds, but all in all, it was good news.

"What if you're mistaken?" asked Hugo's boss.

"I'm not," said Gurney.

By late September, they were still bringing up kimberlite from a depth of 280 metres. Chuck had the core boxes nailed shut and sealed with wax to keep out meddling fingers. They were then quietly flown to Yellowknife, loaded onto a truck and driven non-stop for 48 hours to locked storage in Kelowna. Stu wanted to keep them untouched for a few months. If they examined them and found diamonds, they were legally obligated to announce their results, and this would start a staking rush. But Chuck couldn't wait. When Stu and Marilyn went on vacation to Mexico, Chuck took a 59-kilogram (130-pound) sample from some of the cores and watched as a technician processed it like any routine heavy mineral test, then dumped it into acid that would dissolve anything except diamonds. Chuck took it into his office, locked the door and examined some of the light grit under the microscope. It was breathtaking—there was a diamond. It was just a 1.1-millimetre (0.04-inch) diamond, but it was a diamond nonetheless. He started hunting for others. In a few hours, he had 81 small stones.

Chuck flew to San Francisco to show Hugo. While the potential of a diamond pipe could not be determined from such a small sample, this was great news. If the chairman of BHP was not enthused by the little stones, Hugo was. The existence of one pipe meant there were more. Now Hugo wanted a massive staking push as soon as possible. There was a change in Chuck. The people who had worked for him in the lab and in the field noticed he was even more brusque and tense than usual. Of course, they also surmised that he had not been working on a gold project. Scuttlebutt had it he had hit the big time. Some of them bought Dia Met shares to go with the few hundred penny-stock Dia Met Christmas bonus shares they had received over the years—if they had bothered to keep them. Even Norm, the janitor behind whose name the field camp and company had hidden for years, bought a considerable number of shares.

Dia Met's stock had been trading at 12 cents a share but gradually started edging up. After six months, during which time the stock continued its slow upward climb, the news of the agreement with BHP and the confirmed findings officially went public. A kimberlite pipe with a concentration of 68 carats (13.6 grams) per 100 tons had been discovered buried under 9 metres (30 feet) of glaciated sediment. The low-key announcement was faxed on November 7, 1991, to selected media. It read:

> The following information has been released to the Vancouver Stock Exchange, Canada: the BHP-Dia Met diamond exploration joint venture, in Canada's Northwest Territories, announces that core hole PL 91-1 at Point Lake intersected kimberlite from 455 feet to the end of the hole at 920 feet. A 59 kg sample of the kimberlite yielded 81 small diamonds, all measuring less that 2mm in diameter. Some of the diamonds are gem quality.

Dia Met shares soared as high as $70. The deal with BHP (now BHP Billiton) had succeeded beyond Chuck and Stu's dreams. The announcement touched off the greatest claim-staking rush of the 20th century. Companies worldwide found a new appreciation of Canada and took quick lessons in northern geography and place names: Lac de Gras, in French, and Ekati, in the language of the local Dene, mean Fat Lake. Some Aboriginal elders say the name refers to the fat burned on campfires during the annual fall caribou migrations and hunts of past times. Others claim it is named for the streaks of white quartz through rocks along the shoreline, which resemble bands of fat in the flesh of lake fish or the rumps of migrating caribou during years of plenty. However the name originated, Ekati would become famous as the first diamond mine in Canada.

CHAPTER

8

Diamond Stampede

THERE WERE DIAMONDS IN THE Arctic. Who knew? A Canadian diamond rush began following the announcement of rich diamond-bearing deposits under Point Lake in the Barrens of the Northwest Territories. Dia Met Minerals, which was first listed on the VSE on October 1984, was trading for 50 cents a share in June 1991. Twenty-four months later, the stock hit $66. Chuck Fipke and Stu Blusson had defied conventional scientific wisdom and the statistical odds.

The discovery of diamond-bearing rock in the Northwest Territories marked the beginning of an exploration boom that some compared to the Klondike Gold Rush of 1896–97. When news of the discovery reached the public, investors and prospectors poured billions of dollars into staking claims on any available land and into the development of diamond mines. Over eight million hectares (19,758,430 acres) of permafrost ground was immediately staked out by a multitude of players big and small, domestic and international. Some 260 companies registered claims for 259,000 square kilometres (99,975 square miles) of land. Anybody who had lived in the North and could walk could get hired as a staker. First Nations stakers were in especially high demand.

Out on the tundra, there was a job to be done. As a Dene familiar with the terrain, Yakekan had been hired to stake a claim for a large mining group that held a licence to explore for gemstones underground. The wages for contract stakers were good, but working outside and having the freedom to decide when he would work were equally important. It was a stark, grey-green landscape—nothing but rocks, scrub, lichens and soft, mushy bog surrounding solid ground. Only the constant buzz of swarming insects interrupted the eternal silence. Yakekan knew that when the insects were gone, the cold and snow came.

A helicopter had picked him up at base camp and dropped him here somewhere southeast of Yellowknife. Even though he could live off the land, he had been given a backpack full of supplies as well as stakes and tags to mark off the claim. Yakekan had thought about bringing along snowshoes as the solution to sloshing through muskeg but decided against it. The saving of energy expended in walking would be offset by the bulk. Other professional stakers would also be marking off claims in a rush to grab property. It was expensive to feed and shelter a field crew and to fuel the helicopters, but the exploration companies knew a major mineral discovery came along only once or twice a century, and in a staking rush the last one in only gets the leftovers. The cutthroat competition meant nice guys finished last.

Staking a claim is an act loaded with significance for First Nations people. In 1670, the Hudson's Bay Company became the world's largest property owner when Charles II of England granted them title to the huge area of Rupert's Land—the northern half of present-day Canada—which they advanced by building forts or "factories" across the northwest. Today, a quick trip to the mineral claims or recording office to mark lines on a map, pay a fee and hammer stakes into the ground or cut a line through trees is all it takes to stake a land claim to explore for hidden treasure. Once staked, mineral claims give the holder legal rights to explore and develop the claim and guaranteed ownership of any minerals found there. By comparison, Aboriginal rights in the same territory are far less secure and subject to negotiation with the government.

Local stakers like the man called Yakekan often did not know the identity of the ultimate client. He had not been hired to prospect the area or to provide advice but solely for his considerable knowledge and experience as a staker. His job was to physically put stakes into the ground, a process that gave rise to the term "staking your claim." A claim was a square quarter mile, and stakers were paid by the claim or by the line. But first they had to be certain of their ground location relative to surrounding lakes, rivers, roads and hills, as well as neighbouring claim stakes, and run their boundaries from those claim posts. Today, Global Positioning System (GPS) equipment comes in handy. Just because a map says a post is at a certain location does not always mean that is true. Years ago, prospectors used only a compass and chain to pace off boundaries. Some measurements were not accurate, and disputes were many.

The correct staking procedure is to first find the corners of the intended claim perimeter and sink claim markers with directions right and left of each numbered post or stone cairn. Each claim must have claim tags securely attached to each corner post. The tag is placed on the side facing the next-highest-numbered corner post. Claim boundary lines are marked between posts by regular notching ("blazing") of trees, placing pickets or cutting underbrush.

Today, in areas where the land has been surveyed, map staking is used, which is faster and easier. Claims are marked directly on a map in the record office. Claim dimensions can be much larger, too, with units of 16 to 25 hectares (39.5 to 61.8 acres), and mining companies can have more than one individual stake claims for it, then transfer ownership to the company. But this was not Yakekan's immediate concern. He would finish his current work, be picked up and get more staking jobs. It was better than working in the mines, and he could take time off to hunt and fish. This activity was just the latest intrusion into northern life. First there was the interest in furs, then gold and now diamonds. Who knew what the next economic imperative would be from the white guys?

It seemed everyone had diamond fever and had made a beeline for the Arctic. Who cared if it was winter in the North? Most didn't even know exactly what they were looking for, but the diamond stampede filled Yellowknife with every specialist and labourer who wanted to make a quick buck. Exploration groups spared no expense. After all, they would be as rich as Chuck Fipke when they found diamonds. Of course, De Beers arrived too, which was no surprise to Chuck. Their subsidiary Monopros staked as close to Dia Met's boundaries as possible, their small army of stakers hammering in stakes wherever they could. Later, De Beers would see if there was anything under the ground, but right now it was a land rush, and they just wanted as much territory as possible.

Everybody wanted a piece of the action. Junior mining companies like Aber Resources were there. Even Kennecott, Chuck's first employer, was scrambling for position and lining up joint ventures with other companies. Immense blocks of jealously guarded claims were spreading south from Lac de Gras and north to the Arctic Ocean, and each claim holder was suspicious of his neighbour. As an experienced northern geologist, Chuck had more knowledge than the newbies. His failure to find commercially viable diamonds in the Colorado pipes had been a disappointment, but he remembered how Superior Oil/Minerals had optimistically built a recovery plant on site in advance to separate the diamonds from the kimberlite. With their Colorado mine now proven uneconomical, that plant sat idle. Chuck immediately bought it, and Dia Met, now very solvent, owned the first diamond-processing plant in North America. Investors believed they were set for life.

The diamond discovery was the Cinderella story of North America, considering that Chuck had found the Point Lake kimberlite pipe on the last helicopter flight Dia Met Minerals could afford, and that BHP had uncovered diamond-bearing kimberlite when it drilled the first test hole in September 1991. The frenzy of staking continued into 1992, and Chuck's comment on all the activity was quoted in the media. "We got all the good ground," he stated. "Nobody else is going to find anything." It was dubbed Fipke's Curse. All of Chuck's diamond-hungry competitors took that as a challenge. Norm's Camp was quickly surrounded by other camps housing exploration and sampling teams. They were supplied by aircraft and helicopters bringing in food, equipment and fuel drums. At the peak of the diamond rush, some 45 helicopters and 200 fixed-wing aircraft were buzzing over Lac de Gras. It was pandemonium. Transport Canada issued a safety alert that caution was to be exercised to avoid mid-air collisions.

A helicopter unloads supplies in the Northwest Territories.
R.G. BLACKADER, PHOTO 20289. NATURAL RESOURCES CANADA 2011, COURTESY OF THE GEOLOGICAL SURVEY OF CANADA

But the activity was not all focused on supply missions. So far, only the Point Lake kimberlite pipe was identifiable, and its physical properties could be measured using airborne geophysical surveys. Diamond seekers figured if they could find out what the Point Lake geophysical characteristics were, then they would have a more precise idea of what to look for in their own surveys flown over other areas of interest. Planes and helicopters flew grid patterns over the area with instruments slung under their bellies to measure the physical properties of the rocks below the surface. Hugo Dummett, in charge of BHP's ground oper-ations, realized what was happening and endeavoured to make these geophysical spy missions fail. A large elec-tric cable was strung out around Point Lake and hooked up to a gas generator. When it was turned on, overhead electromagnetic readings were neutralized. A crew was stationed nearby 24-7, and every time they heard an air-craft approach, they switched on the generator. Zap! All the indicator blips flatlined in the snooping planes' instru-ments. The BHP/Dia Met crews found this very funny.

But the opposition soon figured out what was hap-pening. They overflew the area on a random basis—early, late, frequently, infrequently—hoping to catch the crew off guard. Helicopters chartered by junior mining companies came in low over the BHP/Dia Met block claim to scan oper-ations. From the ground, workers observed one helicopter with someone in the passenger seat video-recording them. They recognized the helicopter's markings and realized it belonged to a company owned by the De Beers conglom-erate. Early one Sunday morning, the crew were caught napping when a plane made its first overhead pass. They ran out to start the generator, but it wouldn't catch. They filled it with gas and fired it up, but by then the plane had made two more passes and captured all the data it needed.

Everybody was into the spy game and dirty tricks. BHP also had its own airborne reconnaissance and soon had their own geophysical data on surrounding sites. BHP drills kept company with those from Aber Resources and Monopros, the De Beers affiliate. Barges were assembled so drill rigs could work the most promising sites under northern lakes when the melting ice would no longer hold their weight.

Secrecy from prying airborne espionage was still a priority. Sheds and camp buildings were painted in green, brown and grey tundra camouflage rather than company logos. Drills and fuel drums were hidden under military concealment tarps to disguise the size of operations. In this surreal environment, everybody was hunting for diamonds and everybody knew it, but nobody wanted to admit it. Sampling crews were also under observation. It was important not only to know where companies were drilling but also where they were looking. When the drone of an overhead engine was heard, samplers would pick up their hammers and ore sacks and try to find any available hiding spot until the aircraft was determined not to be hostile.

In the summer of 1992, a few miles from Norm's Camp, BHP drilled nine new sites and hit nine new kimberlite pipes. All nine were diamondiferous, and all nine were under water. Aber Resources drilled 10 holes and found 10 diamondiferous pipes. Nobody knew what De Beers had found, but they and other companies were still there, so they must have hit something. Each new pipe was given a number and a code name to throw off any radio eavesdroppers. BHP's main pipes were tagged Fox, Panda, Koala, Grizzly and Leslie. Analysis showed they could all be profitable mines.

Even though BHP staff did not work directly for Chuck, he still worked them hard for long hours sampling, analyzing, creating software programs, installing hardware and making detailed maps. And his fame was beginning to spread. Media interviewers asked how he had discovered diamonds in the Arctic, but Chuck was not a good interview subject. He often forgot details and substituted other information. Alan Alda, star of the TV program *M*A*S*H*, came to interview him and narrate a segment of the series *Scientific American Frontiers*. But Chuck was uncomfortable when documentaries probed into his life. Journalists called his secretary in Kelowna to ask about him. She provided basic data, like his birthdate: July 22, 1946. She read it off the copy of his driver's licence that she kept handy so it could be easily replaced the many times he misplaced or lost it. His vanity took a hit. The media described him as "balding," with the *Wall Street Journal* calling him "stumpy . . . with muscular forearms." The samplers and other workers whom Chuck had employed took malicious pleasure in referring to him as "Stumpy" when it became clear that after years of benefitting from their labour at minimum wage, he was not going to share his wealth with them. One employee received a Dia Met logo T-shirt, but on his next paycheque there was a deduction for the cost of the shirt. Many had sold their Dia Met stock before its value soared and only recouped a few hundred to a few thousand dollars. They were pissed off.

Stu Blusson's wealth was almost the equal of Chuck's, but he shared none of the publicity, which suited him just fine. Then Stu's mother, now in her 80s, called him. She had seen the Alan Alda documentary about "one man's vision" and knew that her son had been involved in that diamond discovery in the North, but there had been no mention of him on the show. Stu was concerned that his mother, who had never fully supported his decision to become a geologist, could question if his involvement was real or not. His wife, Marilyn, checked through article archives and did not see her husband's name in the stories. She was convinced Chuck was taking all the credit for himself.

Chuck may well have wished for the media spotlight to be turned off him. Articles highlighted his predilection for strip clubs and his blonde Scandinavian assistant, the sole woman among the 31 men at Norm's Camp. They called her Tinkerbell and Buttercup, and the *Financial Post* reported that when Chuck left, the BHP head geologist fired her. When that story broke, Chuck raced madly around Kelowna buying every copy of the magazine so Marlene would not see it. He had forgotten that they had a home subscription to the *Financial Post*, and the magazine would be delivered to their large new residence on the lakeshore. When Chuck made it back home, Marlene was already waiting for him at the door with the story open. Life inside the house remained tense.

Life in the legal arena also heated up. Chuck's secrecy while tracking the diamond-indicator mineral trains and kimberlite pipes came back to bite him. He and Dia Met were sued by a mutual fund that had once held 1.7 million shares of Dia Met but agreed to sell them for 30 cents each before the Point Lake strike. The lawsuit alleged insider trading and suppression of exploration results. Other people who had considered themselves Chuck's partners also sued, claiming Chuck had withheld analyses and investment information from them. Chuck and Stu were sued by Dia Met shareholders, who charged the two had diluted the shares' value and given themselves large interests plus their Dia Met shares in the deal with BHP. Stu was sued by Pioneer Metals, the company whose assignment for him had left Chuck alone on the tundra, which claimed he should have provided an opportunity for them to invest in the project. Like Chuck, Stu's position was that Pioneer Metals made incorrect decisions and wanted in after the fact. Then Stu sued Chuck over a last-minute clause in the BHP deal allowing Stu an interest in the 500,000 buffer claims BHP had staked around the main site if Stu spent money prospecting there himself. Ticked off by his treatment, Stu was insisting that this right be spelled out in great detail and was convinced by his lawyer that the only route was a lawsuit. Chuck was stunned. He was also no longer in charge, even unofficially, on the ground. BHP took over management of the site, and Chuck returned to his CF Minerals lab in Kelowna pending further ore testing and Marlene's wrath.

By 1994, BHP was still convinced there was mine potential in their claims and was proceeding with environmental approvals. On November 5, 1996, the federal government granted approval for mining, and in January 1997, BHP had its regulatory licences in hand. In 1998, BHP opened the Ekati Mine at Lac de Gras, making Dia Met's 29 percent share of the mine worth millions. In 2001, BHP Billiton acquired Dia Met Minerals Ltd. for $687 million to hold an 80 percent ownership in the Ekati Mine. Fipke and Blusson retained 10 percent each. As a result, Chuck and Stu are listed among Canada's 100 wealthiest people—in 2008, Fipke was 67th and Blusson 62nd—each worth around one billion dollars.

9

New Explorations

AFTER DIA MET'S DISCOVERY, a Welsh mining engineer by the name of Grenville ("Gren") Thomas, who had immigrated to Canada in 1964 and immediately found work in Yellowknife, began staking claims southeast of Dia Met's find for his own company, Aber Resources. In Thomas's experience, mines and finds are usually clustered close together, so in a staking stampede it's wise to claim land as close to the initial discovery as possible. Thomas half-jokingly calls this technique "closeology." The terminology may be unusual, but it has proven a useful strategy for junior exploration companies to prospect close to claims held by major mining players. Thomas had staked some 20,234 hectares (50,000 acres) south of

BHP's claim, but much of his "land" was under the water of Lac de Gras.

It was Thomas's daughter, Eira, who proved closeology worked. The diamond fields in the Barrens under the ancient Precambrian rocks and lakes, sand and eskers are home to caribou, bears, wolves and diamond prospectors like Eira, who grew up working summers at her father's uranium exploration camp at Thor Lake near Yellowknife. Eira (pronounced Ira) is the Welsh word for snow, which was falling when she was born. After graduating from the University of Toronto with a bachelor of science in geology in 1990, Eira took a job with a geological consulting company and then took off to see the world. Three months later, she was in South Africa when she heard the news about the huge diamond strike in Canada. Eira flew back home.

In 1994, following in the footsteps of Chuck Fipke and Stu Blusson, she led the Aber Diamond Corporation exploration team that discovered the kimberlite pipes of what would become the Diavik Diamond Mine, Canada's second-largest, located in the Great Slave area some 300 kilometres (186 miles) north of Yellowknife.

Eira had spent the previous winter poring over the fieldwork of other geologists. As the spring exploration teams hit the ground, Eira was dropped by helicopter on the north shore of Lac de Sauvage, accompanied by her husky-shepherd sled dog, Thor, to do as much sampling as she could, as quickly as possible. Thor was supposed to keep her company and protect her from wildlife, but whenever anything wandered by outside the tent at night, he tried to hide in the sleeping bag with her. One day Thor decided to follow a wolf and went off exploring on his own. When he did not return, Eira feared the worst. Wolf packs often lure lone dogs away and attack them. Sadly, she went back to work. But 32 kilometres (20 miles) away, Thor found himself in Norm's Camp, where he met up with Chuck Fipke. When the dog's ownership was confirmed by his tag, Eira got a radio call. No, she could not come and pick him up because Aber staff were not welcome at BHP's drill site, nor did BHP crew have spare time to deliver him to her—such was the competition and secrecy surrounding the diamond rush in the Northwest Territories. Thor was flown all the way back to Yellowknife on the next supply plane and placed in the dog pound. He was reunited with Eira a week later. After his adventure, Thor the timid was nicknamed Thor the Spy Dog by the Thomas family.

In early 1994, as spring was coming to the North and the ice was melting, Eira and her team drilled core samples south of Lac de Gras and sent them off to Ontario for analysis. Large mining exploration companies were drilling all around them as they waited for the results. The results were promising, so they brought in a larger drill. It began to sink in the soft ice; then they hit granite. The drill pipe bent and had to be replaced, but Eira pushed on.

One evening, while the team was examining the day's core samples, they noticed diamond-indicator minerals were present in good concentrations, and they saw an impression of a diamond crystal at the end of one of the cores. But where was the matching core? A frantic search ensued, and when it was finally located, they were looking at an intact diamond crystal. Eira took the core to show her father—her perseverance had paid off. The next step was to go public and get a partner to mine the diamonds.

Eira was immediately dubbed the Queen of Diamonds, a reputation enhanced by her tall, athletic build, pale skin, red hair and intellectual talents as a scientist, strategist, publicist, negotiator and entrepreneur. She became rich as well as famous after Aber Resources Ltd. (now part of Harry Winston Diamond Corporation, which holds a 40 percent interest in the Diavik Diamond Mine) discovered and developed the world-class mine. Eira credits her education for the Diavik find, saying "geochemistry, geophysics and remote-sensing technology all played a role in the discovery."

Production began at Diavik in 2003. The mine consists of three kimberlite pipes on a 20-square-kilometre (7.7-square-mile) island in Lac de Gras, locally called East Island. It is a major source of gems for Harry Winston, known as the "Jeweller to the Stars." Mining giant Rio Tinto, through its subsidiary Diavik Diamond Mines Inc., holds a 60 percent interest in the mine.

Diamond-drill core samples from 122 to 183 metres (400 to 600 feet). J.M. HARRISON. NATURAL RESOURCES CANADA, EARTH SCIENCES SECTOR KGS-542

The groundbreaking father-daughter duo has left Aber, and Gren has started another junior exploration company as well as a Welsh-style pub, the Red Lion, in West Vancouver. Eira is developing the Renard Diamond Project, a deposit 800 kilometres (497 miles) north of Montreal, which will become Quebec's first diamond mine. Now executive chairman and director of Stornoway Diamond Corporation, she partnered with Soquem Inc., a mining and exploration investment company owned by the Quebec government, after it won a hostile takeover bid for Ashton Mining of Canada Inc.

Renard has proven to have three times the ore body of the initial estimates, but raising money in capital markets to put the mine into production is one of many challenges facing the project. Can Eira repeat her Diavik success? She makes it clear that Renard is not another Diavik. Its anticipated production is 1.6 million carats (320,000 grams) a year, versus 8 million carats (1.6 million grams) from Diavik. Renard has one-quarter the gem-quality grade of Diavik, which means it will yield fewer carats per metric ton. The province has committed $130 million in capital funding to build a road to service four mines and a new park in the area. Production at Renard is set to begin in late 2013.

Wearing the Diavik-sourced, brilliant-cut diamond earrings she calls her uniform, as well as other diamond jewellery, Eira recalls looking out over Lac de Gras as a 25-year-old heading the team that made the Diavik find. Her father had cautioned her that although making such a great discovery was an exceptional event, it didn't mean a producing mine would be the result or that she would make more such discoveries in her career. However, Eira recalls that she still felt she "had won the lottery the first time." Now she is out to prove it wasn't just beginner's luck.

Eira wasn't the only young geologist to make her mark in diamond exploration. In 1987, Brad Wood was a second-year geology student at Lakehead University in Thunder Bay when he landed a summer job with De Beers, the international diamond conglomerate. De Beers was looking for diamonds 500 kilometres (310 miles) north of Timmins in the James Bay Lowlands west of Attawapiskat in northern Ontario and had been hiring exploration teams to search for diamonds in the area since the 1950s. Midway through Brad's assignment, the exploration team to which he was assigned had been flown out for some R and R. Brad remained behind to take care of the camp and catch up on some recreational fishing in the nearby river with his father, who had come to keep him company. As they walked along the riverbank one day, they saw a chunk of kimberlite as big as a man's fist lying on top of the ground. Brad knew that the odds were against finding kimberlite, much less finding diamond-bearing kimberlite—about one in a thousand kimberlites actually contain diamonds—and the odds are even smaller of finding a deposit that can become a productive mine. He could hardly wait for the team to return so he could show off his discovery. They were, to put it mildly, extremely excited. Immediate exploration uncovered two kimberlite pipes with a surface area of 15 hectares (37 acres).

Construction of the mine infrastructure started in 2006, and the Victor Mine, the first diamond mine in Ontario, officially opened in July 2008. At Victor, 95 percent of the diamonds are classified as either gem- or near-gem quality. (De Beers' first mine outside Africa, Snap Lake, located northeast of Yellowknife, yields just over 70 percent gem-quality stones.) To maximize local Aboriginal employment, priority is given to the Attawapiskat and other First Nations peoples. The Victor open-pit mine is expected to produce 600,000 carats (120,000 grams) of diamonds annually for 12 years. It has winter road access six weeks a year to bring in equipment and supplies. Personnel are transported to and from the site by air.

Over the years, Brad has been employed by De Beers prospecting in Alberta and the Northwest Territories and working at a South African diamond mine. He is now technical services manager at Victor.

10

Tale of Ekati

IN 1997, STU BLUSSON AND Chuck Fipke received the H.H. "Spud" Huestis Award for Excellence in Prospecting and Mineral Exploration from the Association for Mineral Exploration British Columbia. Like other winners, they were commended for "making a significant contribution, directly or indirectly, to enhance the mineral resources of British Columbia and/or the Yukon Territory, through the original application of prospecting techniques or other geoscience technology." It was just the beginning of commendations and controversy.

When Chuck's girlfriend Tara gave birth to a baby boy, it was the final insult for Marlene. She wanted a divorce, and she wanted nothing less than half of everything she had helped Chuck achieve through her time, support, work and money invested to establish his company and keep it going. Equal treatment was the only acceptable settlement.

The Fipkes separated in 1995, and in 2000, an out-of-court agreement was reached whereby Chuck transferred all his shares in Dia Met Minerals to Marlene. She would receive shares initially worth $123 million. Chuck would retain his 10 percent stake in the Ekati Mine. No other terms of the divorce were disclosed. The official business announcement read:

DIA MET MINERALS ANNOUNCES CHANGE OF MAJOR
SHAREHOLDER.
Kelowna, B.C.—(Business Wire)—Feb. 24, 2000

Dia Met Minerals Ltd. (AMEX:DMM.A) (TSE:DMM.A.) (AMEX:DMM.B) (TSE:DMM.B.) announced today that Mr. Charles Fipke, the Company's founder, Chair and largest shareholder, has agreed to transfer ownership of substantially all of his shares in Dia Met to his estranged wife, Mrs. Marlene Fipke, pursuant to terms of a divorce settlement reached between the parties.

The Company has been informed that the planned transfer involves approximately 1.3 million Dia Met Class A subordinate voting shares, and approximately 5.2 million Dia Met Class B multiple voting shares. Upon completion of the share transfer, Mrs. Fipke will have control over a total of 1.3 million Class A shares and 5.5 million Class B shares. On a fully-diluted basis, these share blocks represent approximately 12.7% of Dia Met's Class A shares; approximately 24.8% of the Company's Class B shares; and approximately 21.0% of Dia Met's total shares outstanding.

"While we recognize the personal and painful nature of this or any divorce settlement, and thus extend our sincere best wishes to both parties, Dia Met is pleased that the distractions and uncertainties caused by this dispute have finally been put to rest," said Mr. James Eccott, Dia Met's President and Chief Executive Officer.

Marlene had more than a passing knowledge of the diamond industry. Since early in their global travels, she had been involved in laboratory analyses and continued to be involved when Chuck established Dia Met and CF Minerals. She understood the significance of the Ekati project and had been responsible for payroll and overseeing Dia Met's laboratory work schedule. At the time, theirs was the largest divorce settlement in Canadian history. "Cost me $200 million," Chuck once said. "Best money I ever spent!"

Nor does Chuck have any regrets about his work, saying, "I did my best. How many people can say that? Everyone thought I was crazy. But most people just never do their best. And I did."

What does a billionaire geologist do after making the greatest gem-quality diamond discovery in North America? For Chuck Fipke, the answer was simple. He followed his two passions: breeding racehorses and undertaking mineral

An aerial view of the Ekati Mine in the Northwest Territories. BHP BILLITON

exploration. In 2003, he paid what at that time was a world-record price of $2.3 million for a yearling—Tale of Ekati, foaled March 31, 2005, which was named after the Ekati Mine. The brown colt won the 2007 Belmont Futurity Stakes and finished fourth in the 2007 Breeder's Cup Juvenile. Chuck said the muddy track at the Juvenile hampered his horse. Apparently he doesn't like to get his feet wet.

In 2008, Chuck started Tale of Ekati in the Kentucky Derby at Churchill Downs. Tale of Ekati had his work cut out for him. While he had won three of his six starts, including the Wood Memorial, a significant Derby preparation race, he was 15 to 1 on the morning line at the Downs. At the age of three, he finished fourth in the Kentucky Derby.

Later, he finished a disappointing sixth in the 2008 Belmont Stakes.

The Ekati mine produces 4 percent of the world's diamonds, some $400 million annually, way above Tale of Ekati's career earnings of $738,000 or the Derby's first place prize of US$1.4 million. But for Chuck, it is the thrill of the chase, and a year or two spent waiting for a winner is a short time for a man who spent 10 years crisscrossing the Arctic tundra hunting for diamonds. He has 12 other horses in Canada training with Roger Attfield, who has won the Queen's Plate seven times. Always one to set goals, Chuck would like to win a Triple Crown and maybe a Breeder's Cup Classic.

The winner's circle is a long way from backcountry British Columbia, where a teenaged Chuck and his pet horse loped across his parents' farm, entered a local country fair race and came in last. Tale of Ekati is now retired to stud, standing for a fee of $15,000. As an owner-breeder, Chuck is shopping for mares to produce another stallion prospect. He owns about 60 mares and is seeking to support Tale of Ekati with 20 to 30 of them. Chuck makes all final breeding decisions, according to his own theories and without commercial consideration. He enjoys the game, beginning with breeding a horse, then racing and continuing the process with other horses from his own successful stock.

Wealth has not slowed Chuck down; instead it has invigorated him with new geological exploration possibilities.

He pursues mining projects in Morocco, Greenland, Angola, Brazil and Canada with the same passion he has for horses. His companies—Cantex Mine Development Corporation and Metalex Ventures Ltd.—are involved in the T1 pipe west of De Beers' Victor Mine in northern Ontario. In Africa, the companies have a licensed project exploring the Sahara in southern Morocco on a 2.6-billion-year-old Archean craton that, as formerly disputed territory, has never been explored. Chuck is in for 60 percent, with the other 40 percent going to the Moroccan government. They are searching for base metals, gold, nickel, copper, cobalt, uranium and, according to Chuck, "We have about 25,000 square kilometres and what we've found is several areas with diamond inclusion indicator minerals, and these old Archean cratons are of course very favourable for diamonds."

They are also exploring in Angola, as Chuck explains in rapid-fire delivery, "We have about 12,000 square kilometres in the headwaters of the Cuango River, the most diamondiferous river in the world. The thing is it's been flown with airborne magnetics and we have about 127 magnetic anomalies, and four of them model at 20 to 39 hectares. We were able during the wet season to test four of the anomalies and all of them were kimberlites."

Research at his CF Minerals lab in Kelowna has kept Chuck at the leading edge of mining technology for decades, but there is still something he would like to know: occasionally there are situations when you get good indicators and no diamonds, and he wants to be able to predict that. It's the same thing in horse racing. Everybody wants to be able to pick the winners.

For years, Stu Blusson drove a beat-up 1979 Ford Mustang with a malfunctioning heater and a cracked window. He was known to park it for free at the Yellowknife Airport, 300 kilometres (186 miles) from the billion-dollar Ekati diamond mine he co-discovered. "I don't really have a sense of being wealthy," he has said. "I'd just as well be living in a tent or sleeping under a spruce tree than in a fancy hotel." Now, he and Marilyn enjoy the wilderness and new discoveries.

Stu, a lower-profile personality than Chuck, has not rested on his post-Ekati laurels either. He is chief executive officer and president of Archon Minerals Limited, a Canada-based company active in the precious metals industry. It explores diamond and other mineral properties in the Northwest Territories in the vicinity of Lac de Gras. Its Buffer Zone Joint Venture Project consists of 85 mineral leases on more than 70,000 hectares (172,975 acres) and the Monument Project includes a lake-based kimberlite pipe.

The Logan Medal is the highest award of the Geological Association of Canada. It is named after Sir William Edmond Logan, the 19th-century Canadian geologist who established the Geological Survey of Canada in 1842 and went on to describe the rocks of the Laurentians in Canada and the Adirondacks in New York State. It is presented annually to an individual for sustained distinguished achievement in Canadian earth science. In 2004, it was presented to Stewart Lynn (Stu) Blusson.

On September 9, 2005, Stu was invested as an Officer in the Order of Canada. The citation reads:

> Renowned for his vision and generosity, Stewart Blusson has been a leader in the fields of mineral exploration and geological research. Following a path carved by ancient glaciers over hundreds of kilometres, he and a partner discovered a diamond deposit in the Northwest Territories. This led to the establishment of the first diamond mine in Canada and has given birth to an industry that is quickly gaining global importance. He has since created a prize for earth and environmental sciences research and made significant gifts to the University of British Columbia and the Sea to Sky University, which will be the country's first private, secular, not-for-profit liberal arts and science institution.
>
> And on October 8, 2009, Simon Fraser University (SFU) conferred on Dr. Stewart Blusson the degree of doctor of science, *honoris causa*, for a life and career dedicated to the sheer joy of discovery. He was described as "a renowned geologist, intrepid prospector, and a visionary philanthropist who has made lasting contributions to his discipline and to higher education."

Epilogue

WHEN STAKING FOR DIAMONDS, THE objective is to claim as much land as possible. When staking for posterity, the objective is to influence the future course of events. Both Stu and Chuck have staked out their place for posterity.

Stu Blusson quietly continues to conduct scientific field-work in the mountains of northern and western Canada and on the Canadian Shield. The multi-millionaire diamond hunter is still looking for new discoveries, and while in town he is not averse to taking the municipal bus or hopping a red-eye flight. In 1998, he donated $50 million to his alma mater, UBC, for innovative research and academic excellence, a gift believed to be the largest single donation ever made to a Canadian public institution by an individual or corporation. In making the donation, he said, "The most important research is often the most basic research, which the public often does not get excited about because it, by itself, is simply another piece of the puzzle. However, its significance will only be recognized later when a different researcher in a distant laboratory builds on this advancement of knowledge to ultimately make a major scientific breakthrough."

Blusson managed to avoid publicity for two years before the media discovered his 2002 donation of $30 million worth of shares in his company Archon Minerals Ltd. to Quest University (formerly Sea to Sky University) in Squamish, BC, the first private, secular, not-for-profit, liberal arts and science university in Canada. In 2008, Blusson Hall at SFU was named to honour the largest private donation the university had ever received, $12 million from Stewart and Marilyn Blusson. "Marilyn and I are interested in creating new approaches for solving big health issues," he says. "Rather than competing with larger research institutions that have medical schools, SFU has strategically focused on research and education aimed at preventing disease, rather than just curing it. This can potentially improve the lives of millions of people around the globe."

Stu secured his spot among global philanthropists by putting up US$10 million to fund the next X Prize. The first X Prize, won by a team of engineers in 2004, launched a high-profile international competition for the first commercial

Stu Blusson, doctor of science, *honoris causa*, gave $12 million to Simon Fraser University.

spacecraft. Stu's Archon X Prize is an international competition to decode the human genome. His prize will go to an individual or corporation that can develop a quick and inexpensive way to sequence a human genome. To collect, the winner must decode 100 human genomes in 10 days.

Chuck, also a UBC alumnus, has been generous too. In 1998, he received an honorary doctorate from Okanagan University College. On November 24, 2008, the new $31.5-million Charles E. Fipke Centre for Innovative Research opened at UBC's Okanagan campus in Kelowna. The environmentally advanced, green-tech facility was made possible through a $26.5-million grant from the BC Ministry of Advanced Education and Labour Market Development, topped up with a $5-million gift from Fipke's foundation. In addition, Chuck contributed $2 million for equipment including LAM-ICPMS, a state-of-the-art mass spectrometer that uses lasers to analyze sample materials. The centre houses a mix of state-of-the-art wet and dry labs, a 300-seat lecture theatre, classrooms and a large glass atrium multipurpose area. In making the donation, Chuck said, "I am honoured and it's a wonderful feeling to be able to give something back to the students and to UBC itself. I'm grateful for the education and the experiences I had as a UBC student."

Chuck is founding board member and a financial donor for WildAid Canada, a conservation organization dedicated to reducing consumer demand for endangered wildlife products and to encouraging responsible energy consumption. With Chuck's financial aid, 12 public-service announcements aired urging Canadians to put an end to bear poaching, among other things. Knowing the value of celebrity,

Chuck Fipke gave $7 million to establish and equip the Charles E. Fipke Centre for Innovative Research at UBC's Okanagan campus in Kelowna. MARGO YACHESHYN. COURTESY OF UBC

he recruited Virgin Airways chairman and tycoon Richard Branson as the organization's first donor and introduced the organization to actress Bo Derek, a supporter whom he met when his horse Silver Charm ran in the 1997 Kentucky Derby. Derek returned as one of Chuck's guests at Churchill Downs when Tale of Ekati ran in the 2008 Kentucky Derby.

Chuck has also been the driving force behind Operation Migration, and since his involvement the whooping crane population of Canada, which had fallen to 29 birds, has now increased to over 500. If he hadn't been a diamond geologist, Chuck says he would have liked to have been an ornithologist, since he has had an interest in birds since his childhood. "We're trying to support the whooping cranes," he says. "I think I'm maybe the principal supporter."

There are many philanthropists giving to worthy causes in Canada, but the discoverers of diamonds under the tundra have chosen to use their wealth to continue the quest for knowledge and encourage others to follow a dream. Chuck Fipke and Stu Blusson have helped to create prestigious new schools, facilitate scientific research, fund university buildings and even breed championship race-horses. Both have made their millions through their own determination and hard work and now dispense them to the causes that have personal meaning to them.

Canadian Diamond Discovery Timeline

1541 French explorer Jacques Cartier believed he had found diamonds at the mouth of Rivière du Cap Rouge, near present-day Quebec City, giving rise to the name Cap Diamant (Cape Diamond). However, the stones he found were just quartz crystals.

1800s Random diamonds have been found in glacial debris in an arc around the Great Lakes from the 19th century to today. It is believed these stones originated in northern kimberlite formations and were moved south by the movement of ancient Ice Age glaciers.

1863 The first Canadian diamond was found buried in glacial debris in Ontario.

1920 A 33-carat (6.6 gram), heavily flawed diamond was found near Peterborough, Ontario, during the construction of a railway track.

1962 Small diamonds were discovered east of Prince Albert, Saskatchewan.

1971 A 0.25-carat (.05 gram), gem-quality diamond was found in glacial sediment near Timmins, Ontario.

1977 Chuck Fipke founded CF Mineral Research and patented the methods that put his laboratory and services into the forefront of geological analysis in North America.

1981 Diamond-indicator minerals were found near the Mackenzie Mountains at the Yukon–Northwest Territories border.

Canadian Diamond Discovery Timeline

1984 Chuck Fipke founded Dia Met Minerals Ltd., which continued to explore in the Northwest Territories until 1991, when the first diamond claims were staked in the Lac de Gras region.

1987 While working for De Beers in the James Bay Lowlands, university student Brad Wood discovered kimberlite formations on an embankment on the Attawapiskat River in northern Ontario. His discovery led to the opening of the Victor Mine in 2008.

1989 As exploration intensified, Fipke found high concentrations of diamond-indicator minerals near Lac de Gras, NWT, suggesting a diamondiferous kimberlite pipe. Fipke started to stake mineral claims.

1990 Dia Met and BHP, an Australian mining conglomerate, signed a joint venture agreement for the Northwest Territories Diamonds Project. BHP agreed to finance exploration costs for shares in any future mine on the property. BHP would own 51 percent, Dia Met would hold 29 percent and Fipke and Blusson would have 10 percent each.

1991 Dia Met and BHP's discovery of diamonds at Point Lake sparked the NWT Diamond Rush, the greatest staking rush since the Klondike Gold Rush of 1896–97.

1992 Chuck Fipke was named Mining Man of the Year by the *Northern Miner* in recognition of his technical skills, years of hard work and dedication to a single goal.

1993 BHP opened an office in Yellowknife and began winter drilling at the sites of the Fox, Leslie and Koala kimberlite pipes, with bulk sampling at Fox.

Aber Resources Ltd. and Kennecott Canada Exploration opened a Diavik Project office in Yellowknife.

1994 BHP submitted a full-scale mining project proposal for review by the Northwest Territories Regional Environmental Review Committee (RERC). The federal minister of Indian and Northern Affairs recommended the Northwest Territories Diamonds Project undergo a public environmental assessment under the Environmental Assessment Review Process Guideline Order (EARPGO). BHP undertook more winter drilling at Panda, Koala, Fox, Leslie and Misery kimberlite pipes, with bulk sampling at Panda.

1995 BHP submitted its Environmental Impact Statement (EIS) on the Ekati Diamond Mine to the federal Environmental Assessment and Review Process (EARP).

1996 Chuck Fipke founded Cantex Mine Development Corporation.

The EARP recommended the "Government of Canada approve the Diamonds Project."

An Impacts and Benefits Agreement (IBA) was reached with the Dogrib Treaty 11 Council.

A Socio-Economic Agreement was signed with the government of the Northwest Territories.

Diavik Diamond Mines Inc. was created, with a head office in Yellowknife.

1998 On October 14, Dia Met Minerals Ltd. and BHP Diamonds Inc. announced the official opening of Canada's first diamond mine: the Ekati Diamond Mine.

1999 Ekati produced its first one million carats of diamonds.

2000 De Beers Canada acquired Winspear Resources and with it the Snap Lake diamond deposit.

The NWT government certified the first Canadian Arctic™ diamond and began to promote the product as well as the secondary diamond-cutting and polishing industry in the North.

2001 BHP Limited merged with Billiton Plc to form BHP Billiton, which then acquired Dia Met Minerals Ltd. for $687 million. BHP Billiton now owned 80 percent of the Ekati Mine; Fipke and Blusson each retained 10 percent.

The construction of the Diavik Diamond Mine began.

Canadian Diamond Discovery Timeline

2003 The North's second diamond mine, Diavik Diamond Mine, owned by a joint venture partnership between the Harry Winston Diamond Corporation and Diavik Diamond Mines Inc., a subsidiary of the Rio Tinto Group, began production near Ekati.

BHP launched its Canada Mark™ diamond certification.

2004 De Beers opened its underground Snap Lake Mine, northeast of Yellowknife, its first mine outside Africa.

2005 Stewart Lynn "Stu" Blusson, O.C., Ph.D., invested as an Officer of the Order of Canada.

2006 De Beers began construction of its open-pit Victor Mine in the James Bay Lowlands of northern Ontario, its second diamond mine in Canada and the first in Ontario.

Stu Blusson, in addition to giving millions to Canadian universities, donated the largest medical prize in history, US$10 million, for the Archon X Prize named after the ancient Archean Craton, or Archon, in Canada, where diamonds were discovered. The prize will be awarded to the person or group developing a quick (100 people in 10 days) and inexpensive way to sequence a human genome.

2008 De Beers' second Canadian diamond mine, the Victor Mine, opened in northern Ontario.

2009 The $31.5-million Charles E. Fipke Centre for Innovative Research opened at UBC's Okanagan campus. The centre was funded by a $5-million gift from Chuck Fipke (plus $2 million for equipment) and money from the BC Ministry of Advanced Education and Labour Market Development.

Glossary

Airborne survey A geophysical survey conducted by helicopter or fixed-wing aircraft to identify properties of rock below the surface, such as magnetic state, specific gravity, radioactivity, etc.

Alluvial diamonds Diamonds washed from their location of origin, usually from kimberlite deposits, into deposits of clay, silt and sand left by flowing water along a riverbank, shoreline or seabed. Industrial alluvial mining removes the overburden (covering soil) to find diamonds, while marine mining excavates diamonds that have been transported to the ocean floor. The world's largest known gem-quality alluvial diamond deposit is located along the Namib Desert coastline of southwestern Africa. Many alluvial diamond deposits occurred in the Pleistocene (1.8 million to 10,000 years ago) and Holocene (the last 10,000 years of earth's history) epochs.

Archaean Eon Refers to an early part of Precambrian time (before or about 2,500 million years ago), when the earth was much hotter than today and the oldest-known rocks and fossils were formed.

Barrens (also Barren Lands, Barren Grounds) A 130,000-hectare (321,235-acre) area of tundra at the top of North America at the Arctic Circle that contains some of the oldest known rocks in the world. These 4-billion-year-old formations are a likely source of diamonds and other minerals such as zinc, copper and gold. The vast Barrens are home to the Dene Nation and the Inuit, who previously lived by hunting the herds of caribou as well as wolves, bears and other wildlife. Many Dene and Inuit people now work in the mining industry.

Glossary

Bort (also Boart) Industrial-grade shards of gem-quality diamonds or imperfect diamonds that are used for tools such as drill bits and abrasive grit, rather than jewellery.

Carat A standard unit to measure the weight of a diamond. It is thought to be derived from carob, uniformly sized seeds used to balance ancient scales for gemstones. More seeds indicated a heavier and more valuable stone. One carat equals 100 points. Metrically, one carat is equal to 200 milligrams (one-fifth of a gram) so five carats equal one gram. On the imperial side, one carat is equal to .007 ounces, therefore 141.7 carats equal one ounce. "Carat" should not be confused with "karat," which is used to describe the amount of pure gold in an object.

Carbon A non-metallic element found free in nature in three structural (allotropic) forms, one of which is diamond.

Cartel Producers of a product that join together to control production, sale and price, thus forming a monopoly that restricts competition in that commodity or industry. We refer to a Middle East oil cartel and to De Beers Group, which operates a diamond cartel.

Certified When a gemological laboratory has certified the quality of a diamond (clarity, color, carat weight, cut and proportion), it issues a certificate, in the form of a laminated document, describing these characteristics in detail.

Claim A portion of land staked out and registered according to federal or local laws by a claimant or an organization.

Conflict diamonds Also called "blood diamonds," these are diamonds used to finance wars, terrorist groups or other illegal activities.

Core A cylindrical section of rock, usually about an inch in diameter, drilled out and brought to the surface for geological examination and laboratory analysis.

Craton An old and geologically stable portion of the earth's crust. Most diamondiferous kimberlites are associated with Archean cratons.

The Slave craton in the Northwest Territories lies from Great Slave Lake to Coronation Gulf on the Arctic Ocean and is one of the oldest rock formations on earth.

De Beers Group A large international organization that has controlled most of the production and distribution of the global diamond market since its founding in 1888 by Cecil Rhodes. Some call it a monopoly. The De Beers Group is based in Johannesburg, South Africa, and London, England.

Dene The Dene Nation (Dene means "the people") is part of the Athapaskan linguistic group, the largest Native linguistic group in North America. The Dene Nation has lived in central and northwestern Canada from the Mackenzie Delta, west into Alaska, east into Nunavut and south into the prairies. Their homeland is referred to as Denendeh, meaning "the Creator's spirit flows through this land."

Diamond (gem-quality) Rough, gem-quality diamonds are sent to one of the main diamond-cutting and trading centres: Antwerp, Mumbai, Tel Aviv, New York, China, Thailand or Johannesburg. There, experts known as diamantaires cut the rough stones into round brilliant, oval, pear, heart and emerald shapes and then polish them. The diamonds are classified by their cut, colour, clarity and carat weight (known as the Four Cs) and sold to diamond wholesalers or jewellery manufacturers in one of 24 registered diamond exchanges, called bourses, around the world. There are natural diamonds and synthetic diamonds.

Diamond (industrial-grade) The industrial-grade market is much different from the gem-quality one. Industrial diamonds are valued for hardness and heat conductivity, so gem qualities like clarity and colour are not important. Some 80 percent of mined diamonds, approximately 100 million carats, or 20,000 kilograms annually, are not suitable for gem use but still have industrial value. Man-made (synthetic) diamonds were also put to industrial use after their invention in the 1950s, and 3 billion carats, or 600 metric tons, are produced annually. Industrial diamonds are used in machinery for cutting, drilling, grinding and polishing.

Diamonds are embedded into drill bits or saw blades, or ground into powder for grinding and polishing. Other uses are in high-pressure laboratory experiments, high-performance bearings and specialized windows that do not absorb infra-red radiation, used in NASA space probes.

Diamond brands Soon after the Ekati mine began producing, BHP Billiton sold rough diamonds to Northwest Territories manufacturers, and Canadian brands started to appear. This successful venture inspired many others to brand by Canadian origin: Polar Bear, Polar Ice, Eskimo Arctic Ice, Maple Leaf, Aurias, Igloo, Canadian Arctic and Tundra are some symbols of Canadian diamonds. Under the Canadian Competition Bureau's guidelines, stones sold as Canadian must have been mined in Canada. The industry established the Voluntary Code of Conduct for Authenticating Canadian Diamond Claims, which provides a tracking procedure.

Diamondiferous Rock or alluvial material containing diamonds.

Ekati Canada's first diamond mine, owned by BHP Billiton, produces 4 percent of the world's diamonds by volume and 6 percent by value. Ekati, near Lac de Gras in the Northwest Territories, is about 300 kilometres (186 miles) northeast of Yellowknife. It is a joint venture between BHP Billiton Diamonds Inc. (80 percent) and geologists Charles E. Fipke and Dr. Stewart E. Blusson (10 percent each) after BHP Billiton bought out Dia Met Minerals Ltd.'s holdings in mid-2001. Environmental water use, waste management, land use and rehabilitation are monitored. Some 314 hectares (776 acres) of tundra have been used for mine construction; 611 hectares (1,510 acres) of the total lease area of 10,960 hectares (27,082 acres) have been affected by operations.

Erratics Boulders, rocks or pebbles carried by prehistoric glacial ice and often deposited hundreds of kilometres from their place of origin. They differ from the size and type of rock native to the area in which they have come to rest.

Esker Winding ridges of gravel and sand left behind on the surface by melting water from retreating glaciers or ice sheets. Their size varies, but they can be many kilometres long. Roads can be built along eskers to reduce costs, such as the Denali Highway in Alaska. Two of the largest Canadian eskers are the Thelon esker in the Northwest Territories and Nunavut (800 kilometres/497 miles long) and the Munro esker near Munro Lake in northern Ontario (250 kilometres/155 miles long and 5 kilometres/3 miles wide).

Exploration The work involved in looking for valuable deposits of minerals. It can include surface and underground investigations, as well as geological reconnaissance involving remote sensing, photogeology, geophysical and geochemical methods.

Facies In geology, facies are bodies of rock with specified, distinct characteristics formed under certain conditions of sedimentation and are thus reflective of a particular process or environment.

Four Cs Cut, Clarity, Colour, Carat: The brilliance or "fire" of a diamond depends on cutting and polishing the facets to allow the maximum amount of light that enters through its top to be reflected and dispersed back; the visibility, number and size of internal flaws or inclusions that occur during the formation process determine the clarity (clear diamonds create more brilliance and are more highly prized and priced); colourless diamonds are the most desirable, since they allow the most refraction of light; a carat is the unit of weight by which a diamond is measured, and less-common large diamonds are more expensive than smaller ones. There is also a fifth C: certificate of origin and grading, which evaluates a diamond and its source.

Geology The science concerned with the study of the solid earth and the rocks of which it is composed and the processes by which is has been changed or is now being changed.

Geophysics The measurement of the distinct physical properties of rock, such as specific gravity, magnetism, electrical and seismic conductivity and radioactivity, using specialized measuring devices and computers.

Glossary

Geophysical survey A scientific method of prospecting that measures the physical properties of rock formations. Unique area geophysical properties and relationships are mapped by one or more methods. Surveys are helpful in diamond exploration because kimberlites often have distinctive geophysical signatures compared to surrounding rocks. Other applications are in science, archaeology and engineering for both industrial purposes and academic research.

Glacial drift Sediment (rock material) that is either now in transport in glaciers or has been deposited by glaciers or rafted by icebergs.

Grade The concentration of metal or valuable mineral in a body of rock. In diamond exploration, grade is determined by the number of carats in a physical unit of ore, usually expressed in carats per ton, which denotes how rich a find may be.

Indicator-mineral train The area where kimberlite indicator minerals are concentrated is variously described as indicator-mineral train, fan, dispersion or anomaly. The train was created by glacial ice carrying surface material dispersed over vast areas. Kimberlites are soft and tend to weather more than the surrounding rocks, so any kimberlite that lay in the path of a glacier was scoured out, incorporated into the glacial drift and dispersed over the landscape. Kimberlite indicator minerals were dispersed the same way.

Kimberley Process Certification Scheme (KPCS) This agreement imposes requirements to certify rough diamonds as conflict-free. As of December 2009, the KPCS had 49 members in 75 countries (the European Union counted as one). The Kimberley Process is a joint government, industry and civil-society initiative to try to control conflict diamonds used to finance wars. It is not an independent international body but a coalition of countries; therefore it has difficulty enforcing its own policies. Trade in illicit diamonds has funded slaughters in Angola, Congo, Côte d'Ivoire and Sierra Leone.

Kimberlite A rare igneous-rock matrix composed of diamond-indicator minerals like calcium carbonate, garnet, olivine, phlogopite, pyroxene, serpentine and upper mantle rock, with a variety of trace minerals. Kimberlite is the most common host deposit for diamonds. Most kimberlite is called blue-ground or yellow-ground kimberlite and is found worldwide. The name kimberlite is derived from the South African town of Kimberley where the first diamonds were found in this type of rock. When kimberlite is found, it is necessary to evaluate its diamond content: concentration (carats per ton), the size of the deposit and the size and quality of the diamonds. Kimberlites occur on every continent. According to the *Journal of Geochemical Exploration*, the total number of known primary host-rock occurrences is generally accepted as 5,000, of which 500 are diamondiferous, 50 have been or are being mined and 15 are large, active mines. In most of these occurrences, diamonds are present in concentrations of less than one part in five million.

Kimberlite pipes Kimberlite occurs in the earth's crust in vertical structures known as kimberlite pipes, the most significant source of diamonds. But only one in every 200 kimberlite pipes contains gem-quality diamonds. Diamonds form at a depth greater than 150 kilometres (93 miles) beneath the earth's surface. After their formation, diamonds are carried to the surface of the earth by volcanic activity. A mixture of magma (molten rock), minerals, rock fragments and occasionally diamonds forms pipes shaped like champagne flutes or carrots as they approach the earth's surface. These pipes are called kimberlites. Kimberlite pipes can lie directly underneath shallow lakes formed in inactive volcanic calderas or craters. Lac de Gras kimberlites are similar to those of South Africa and Russia, with all the kimberlite pipes covered by small lakes.

Laser identification The process of inscribing diamonds with either a special logo or general information about brand name, number, etc.

Lamproite pipes Lamproite pipes also produce diamonds, but to a lesser degree than kimberlite pipes. Lamproite pipes are created in a similar way to kimberlite pipes, except that boiling water and volatile compounds in the magma acted corrosively on the overlying rock, resulting in a broader cone of hollowed-out rock at the surface. This results in a martini-glass–shaped deposit as opposed to kimberlite's champagne-flute or carrot shape.

Lode A mineral deposit in solid rock that can be imbedded in a crack or form a vein between layers of rock.

Maple Leaf Diamonds Mined at Ekati, each of these diamonds bears a distinct maple-leaf laser mark and serial number not visible to the naked eye to ensure authenticity from rough stone to finished gem.

Open-pit mining Open-pit mining, or open-cast mining, is a method of extracting rock or minerals from the earth by means of an open hole or burrow. The method is used when deposits of minerals are found near the surface (or, when mining diamonds, along kimberlite pipes), when the overburden or surface material covering the deposit is relatively thin, or when the minerals are imbedded in structurally unstable earth like cinder sand or gravel that is not suitable for tunnelling underground. Pit lakes tend to form at the bottom of open pit mines as groundwater and rain collect there.

Pan To wash gravel, sand or crushed rock samples in order to isolate gold or other valuable minerals by their higher density.

Placer A deposit of sand and gravel containing valuable minerals such as gold, tin or diamonds. Placer diamond mining, also known as sand-bank mining, is used to extract diamonds and minerals from alluvial deposits on the surface, often on riverbanks, without tunnelling. Excavation uses water pressure, also called hydraulic mining, or surface excavating equipment.

Polar Bear Diamonds These diamonds are mined, cut and polished in a factory located in Yellowknife, NWT, that employs 100 percent Canadian workers. In 2000, the Government of the Northwest Territories (GNWT) instituted an in-house certification process under the brand Government Certified Canadian Diamond to authenticate diamonds that have been mined, cut and polished in the NWT. An official certificate issued by the GNWT accompanies each certified diamond with individual information about that diamond. Only the GNWT issues these certificates. This statement is important because there is no commercially available technology that can test a polished diamond to prove where it was mined. Other certifications confirm that the diamond was mined in Canada but could have been cut and polished elsewhere. Like a fingerprint, every polished diamond has a unique visual signature. No two are alike. The GNWT scans its polished diamonds using GemPrint™ and stores the images in a confidential centralized database.

Precambrian Shield The Precambrian Shield is an extensive structure of the earth's crust composed of exposed rocks formed during the Archean or Proterozoic eons, which comprise the Precambrian Era, which ended 544 million years ago. Shield rocks are among the oldest and most stable on earth. The Precambrian mountains have since eroded away, creating the low, rolling rock plain of today. The best-known examples are the Canadian Shield and the Baltic Shield in Scandinavia.

Sample A small portion of rock or mineral deposit collected and taken away so the composition and quality can be analyzed in a laboratory. A mini-bulk sample is typically one to one hundred tons of kimberlite and associated indicator minerals collected as part of an initial step in developing a viable diamond mine. A bulk sample is a large amount of mineralized rock, frequently hundreds of tons, selected as representative of the ore body being sampled for metallurgical characteristics.

Staking The measuring of an area of ground and marking its outline with stakes or posts to establish and acquire mineral rights.

Bibliography

Bielawski, Ellen. *Rogue Diamonds: Northern Riches on Dene Land*. Seattle: University of Washington Press, 2003.

Canadian Arctic Resources Committee. *Environmental Assessment Guidelines for the Completion of a Comprehensive Study of Proposed Diavik Diamonds*. Ottawa: Natural Resources Canada, 1998.

Even-Zohar, Chaim. *From Mine to Mistress: Corporate Strategies and Government Policies in the International Diamond Industry*. London: Mining Communications Ltd., 2007.

Frolick, Vernon. *Fire Into Ice: Charles Fipke and the Great Diamond Hunt*. Vancouver: Raincoast Books, 1999.

Hart, Matthew. *Diamond: A Journey to the Heart of an Obsession*. New York: Walker & Company, 2001.

Houston, James. *Black Diamonds: A Search for Arctic Treasure*. Toronto: McClelland and Stewart, 1982.

Kanfer, Stefan. *The Last Empire: De Beers, Diamonds and the World*. New York: Farrar, Straus and Giroux, 1995.

Krajick, Kevin. *Barren Lands: An Epic Search for Diamonds in the North American Arctic*. New York: Times Books, 2001.

Le Cheminant, A.N, D.G. Richardson, R.N.W. DiLabio, and K.A. Richardson, eds. *Searching for Diamonds in Canada*. Ottawa: Geological Survey of Canada, 1996.

Moon, Charles J., Michael Whateley, and Anthony M. Evans, eds. *Introduction to Mineral Exploration*. Toronto: Wiley-Blackwell Publishing, 2006.

Zoellner, Tom. *The Heartless Stone: A Journey through the World of Diamonds, Deceit, and Desire*. New York: St. Martins Press, 2006.

Index

Aber Resources Ltd., 79, 83, 88–92, 111
Archean, 42, 47, 101, 113, 115, 122
Archon X Prize, 106, 113
Arctic Circle, 54, 69, 114
Attawapiskat, 94, 111

Barrens (Barren Lands), 8, 12, 24, 27, 45, 50, 51–52, 57, 62, 75, 89, 114
BHP Billiton, 73, 87, 99, 112, 117. *See also* Broken Hill Proprietary (BHP)
Blackwater Group, 40, 43
Blackwater Lake, 38, 41, 46, 54, 57, 63
Blusson, Marilyn (née Ballantyne), 36, 72, 85, 102, 105
Blusson, Stewart (Stu), 12, 22–28, 35–50, 51–52, 54, 56, 60, 68, 69, 71–72, 73, 75, 85, 86, 87, 89, 96, 102–6, 109, 111, 112, 113, 117
Broken Hill Proprietary (BHP), 31, 68, 69, 72, 73, 80, 82–84, 85, 86–87, 90, 111, 112, 113, 117

Cantex Mine Development Corporation, 101, 112
Cap Diamant, 7, 110
Carmack, George Washington, 8
Cartier, Jacques, 7, 110
CF Mineral Research Ltd., 32, 33, 39, 40, 50, 54, 55, 61, 86, 98, 101, 110
Clifford's Rule, 46–47
Cominco, 31, 32, 34
Coppermine River, 8, 53

De Beers, 10, 23, 31, 37, 39–40, 41, 46, 54, 58, 69, 79, 82, 83, 93–95, 101, 111, 112, 113, 115, 116
Dene, 62, 73, 76, 114, 116
Dia Met Minerals Ltd., 49, 52, 56, 58, 60, 61, 62, 63, 68, 69, 70, 72, 73, 75, 80, 82, 84, 86, 87, 88, 97–98, 111, 112, 117
diamonds
conflict (blood), 10–11, 115

formation of, 6, 8, 13, 35, 119, 120, 121
shapes of, 11, 12
Diavik Diamond Mine, 89, 91, 93, 111, 112, 113
Dogrib, 62, 112,
Dummett, Hugo, 37, 39, 40, 41, 45, 47, 48, 50, 68, 69, 71, 72, 82

Ekati Diamond Mine, 74, 87, 97, 98, 99, 100, 102, 103, 112, 113, 117, 121
esker, 5, 6, 45, 51, 52, 53–54, 55, 56, 57, 59, 60, 63, 65, 66, 89, 117–18
Exeter Lake, 61, 65, 70

Fipke, Charles (Chuck), 12, 14–21, 25–28, 29–35, 36–41, 44–50, 51–60, 61–73, 75, 79–80, 84–87, 90, 96–102, 104, 107–9, 110, 111, 112, 113, 117
Fipke, Mark, 18, 21, 29, 33, 48, 58, 64–67
Fipke, Marlene (née Pyett), 16–19, 21, 29, 30, 32, 36, 40, 48, 85, 86, 96–98
Fipke, Wayne, 14–15, 32, 49, 56, 60
Frobisher, Martin, 7

G–10 (garnets), 34, 59
Geological Survey of Canada (GSC), 19, 23–24, 27, 28, 36, 38, 39, 42, 51, 52, 102
Glacial Divide, 54
Glacial Lake McConnell, 47
Great Slave Lake, 8, 54, 115
Gurney, John, 34, 35, 36–37, 67, 68, 71

Harry Winston Diamond Corporation, 91, 113
Hearne, Samuel, 8
H. H. "Spud" Huestis Award for Excellence in Prospecting and Mineral Exploration, 96
Kelowna, 14, 18, 32, 33, 36, 39, 41, 44, 45, 55, 61, 62, 67, 71, 84, 85, 97, 101, 107, 108

Index

Kennecott Canada Exploration, 111
Kennecott Copper Corporation, 21, 29, 79
Kimberley Process Certification Scheme
(KPCS), 10–11, 119
kimberlite, 6, 8, 9, 13, 31, 32, 34–35, 36, 38, 39,
42, 46, 57, 58, 66, 67, 69, 71, 73, 80, 81–83, 86,
89, 91, 94, 101, 102, 110, 111, 112, 114, 115,
119, 120, 121, 122
Klondike Gold Rush, 75, 111

Lac de Gras, 9, 43, 52, 53, 55, 57, 59, 63, 65, 67,
70, 73, 79, 80, 87, 89, 90, 91, 93, 102, 111,
117, 120,
Lac de Sauvage, 89
Lac la Martre, 46, 47, 54
Logan Medal, 102

Mackenzie, Dave, 49, 56, 59, 60, 63–64, 65,
66, 71
Mackenzie Mountains, 19, 20, 28, 42, 48, 57, 110
Metalex Ventures Ltd., 101
Monopros, 79, 83
Mountain Diatreme 38, 48, 57

New Guinea, 21, 29
Norman Wells, 19, 20, 39, 44, 46, 57
Norm's Camp, 61, 64, 70, 71, 80, 83, 85, 90
Norm's Manufacturing, 54, 65

Okanagan University College, 107
Operation Migration, 108–9
Order of Canada, 103, 113

Point Lake, 67, 69, 70, 73, 75, 80–82, 86, 111
Pointe de Misère, 65, 67

Quest University, 105

Rhodes, Cecil, 10, 116

sampling, 35, 45, 46, 50, 53, 54, 57, 59, 67, 80, 83,
84, 89, 111, 112
Sayunei Range, 24, 38
Simon Fraser University (SFU), 103, 105, 106
Snap Lake, 94, 112, 113
staking, 8, 57, 59, 61–63, 64, 65, 66, 68, 69, 72,
73, 76–79, 80, 88, 104, 111, 122
Stornoway Diamond Corp., 92
Superior Oil Minerals Division, 33, 34, 35, 36,
37, 38, 39, 40, 46, 50, 69, 80

Tale of Ekati, 99–100, 108
Thomas, Eira, 89–93
Thomas, Grenville (Gren), 88, 89, 91, 92, 93
Thor, 89, 90

University of British Columbia (UBC), 17, 18,
19, 20, 23, 33, 103, 104, 107, 108, 113

Vancouver Stock Exchange (VSE), 49, 73, 75
Victor Mine, 94, 101, 111, 113

WildAid Canada, 107
Wood, Brad, 93–95, 111

X Prize, 105, 106, 113

Yakekan, 76–78
Yellowknife, 40, 46, 52, 53, 57, 59, 61, 62, 67, 69,
71, 76, 79, 88, 89, 90, 94, 102, 111, 112, 113,
117, 121

Acknowledgements

No work is ever created in isolation. There are numerous inputs, influences, interpretations, experiences and events that flavour the finished product. I especially appreciate the research help provided by Bob Mercier and access to his extensive library. Living in a resource-rich environment like the National Capital Area, with its museums, universities and libraries, plus the federal Library and Archives Canada, is an experience any taxpaying writer would savour.

I would like to thank Lesley Reynolds for her editing expertise, as well as Martin Legault, photo collection coordinator at the Natural Resources Canada Library, and Joanne Tremblay at the Earth Sciences Sector Copyright Co-ordination Office, Natural Resources Canada, for their assistance in sourcing illustrations.

About the Author

L.D. Cross is an Ottawa writer and member of the Professional Writers Association of Canada and the Canadian Authors Association. Her business and lifestyle articles have appeared in Canada and the United States. Her publication credits include magazines such as *WeddingBells, Home Business Report, Legion Magazine, Profit Magazine, enRoute, AmericanStyle, Fifty-Five Plus, Health Naturally, Antiques!, Airborn* and *This Country Canada,* as well as the *Globe and Mail* newspaper. Her creative non-fiction has been recognized by the International Association of Business Communicators, Ottawa Chapter, EXCEL Awards for features and editorial writing as well as the National Mature Media Awards for her writing about seniors.

She is the author of other books in the Amazing Stories series, including *Ottawa Titans: Fortune and Fame in the Early Days of Canada's Capital; Spies in Our Midst: The Incredible Story of Igor Gouzenko, Cold-War Spy; The Quest for the Northwest Passage: Exploring the Route Through Canada's Arctic Waters;* and *The Underground Railroad: The Long Journey to Freedom in Canada.* She is also a co-author of *Inside Outside: In Conversation with a Doctor and a Clothing Designer* and *Marriage is a Business.*

More Great Books in the Amazing Stories Series

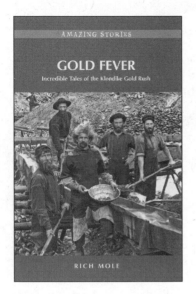

Gold Fever

Incredible Tales of the
Klondike Gold Rush

Rich Mole

(ISBN 978-1-894974-69-1)

In 1897, tens of thousands of would-be prospectors flooded into the Yukon in search of instant wealth during the Klondike Gold Rush. In this historical tale of mayhem and obsession, characters like prospectors George Carmack and Skookum Jim, Skagway gangster Soapy Smith and Mountie Sam Steele come to life. Enduring savage weather, unforgiving terrain, violence and starvation, a lucky few made their fortune, and some just as quickly lost it. The lure of the North is still irresistible in this exciting account of a fabled era of Canadian history.

Visit www.heritagehouse.ca to see the entire list of books in this series.